Graduate Texts in Mathematics **203**

T0280307

Springer
New York
Berlin
Heidelberg
Barcelona
Hong Kong
London
Milan
Paris
Singapore
Tokyo

Graduate Texts in Mathematics

1 TAKEUTI/ZARING. Introduction to Axiomatic Set Theory. 2nd ed.
2 OXTOBY. Measure and Category. 2nd ed.
3 SCHAEFER. Topological Vector Spaces. 2nd ed.
4 HILTON/STAMMBACH. A Course in Homological Algebra. 2nd ed.
5 MAC LANE. Categories for the Working Mathematician. 2nd ed.
6 HUGHES/PIPER. Projective Planes.
7 SERRE. A Course in Arithmetic.
8 TAKEUTI/ZARING. Axiomatic Set Theory.
9 HUMPHREYS. Introduction to Lie Algebras and Representation Theory.
10 COHEN. A Course in Simple Homotopy Theory.
11 CONWAY. Functions of One Complex Variable I. 2nd ed.
12 BEALS. Advanced Mathematical Analysis.
13 ANDERSON/FULLER. Rings and Categories of Modules. 2nd ed.
14 GOLUBITSKY/GUILLEMIN. Stable Mappings and Their Singularities.
15 BERBERIAN. Lectures in Functional Analysis and Operator Theory.
16 WINTER. The Structure of Fields.
17 ROSENBLATT. Random Processes. 2nd ed.
18 HALMOS. Measure Theory.
19 HALMOS. A Hilbert Space Problem Book. 2nd ed.
20 HUSEMOLLER. Fibre Bundles. 3rd ed.
21 HUMPHREYS. Linear Algebraic Groups.
22 BARNES/MACK. An Algebraic Introduction to Mathematical Logic.
23 GREUB. Linear Algebra. 4th ed.
24 HOLMES. Geometric Functional Analysis and Its Applications.
25 HEWITT/STROMBERG. Real and Abstract Analysis.
26 MANES. Algebraic Theories.
27 KELLEY. General Topology.
28 ZARISKI/SAMUEL. Commutative Algebra. Vol.I.
29 ZARISKI/SAMUEL. Commutative Algebra. Vol.II.
30 JACOBSON. Lectures in Abstract Algebra I. Basic Concepts.
31 JACOBSON. Lectures in Abstract Algebra II. Linear Algebra.
32 JACOBSON. Lectures in Abstract Algebra III. Theory of Fields and Galois Theory.
33 HIRSCH. Differential Topology.
34 SPITZER. Principles of Random Walk. 2nd ed.

35 ALEXANDER/WERMER. Several Complex Variables and Banach Algebras. 3rd ed.
36 KELLEY/NAMIOKA et al. Linear Topological Spaces.
37 MONK. Mathematical Logic.
38 GRAUERT/FRITZSCHE. Several Complex Variables.
39 ARVESON. An Invitation to C^*-Algebras.
40 KEMENY/SNELL/KNAPP. Denumerable Markov Chains. 2nd ed.
41 APOSTOL. Modular Functions and Dirichlet Series in Number Theory. 2nd ed.
42 SERRE. Linear Representations of Finite Groups.
43 GILLMAN/JERISON. Rings of Continuous Functions.
44 KENDIG. Elementary Algebraic Geometry.
45 LOÈVE. Probability Theory I. 4th ed.
46 LOÈVE. Probability Theory II. 4th ed.
47 MOISE. Geometric Topology in Dimensions 2 and 3.
48 SACHS/WU. General Relativity for Mathematicians.
49 GRUENBERG/WEIR. Linear Geometry. 2nd ed.
50 EDWARDS. Fermat's Last Theorem.
51 KLINGENBERG. A Course in Differential Geometry.
52 HARTSHORNE. Algebraic Geometry.
53 MANIN. A Course in Mathematical Logic.
54 GRAVER/WATKINS. Combinatorics with Emphasis on the Theory of Graphs.
55 BROWN/PEARCY. Introduction to Operator Theory I: Elements of Functional Analysis.
56 MASSEY. Algebraic Topology: An Introduction.
57 CROWELL/FOX. Introduction to Knot Theory.
58 KOBLITZ. p-adic Numbers, p-adic Analysis, and Zeta-Functions. 2nd ed.
59 LANG. Cyclotomic Fields.
60 ARNOLD. Mathematical Methods in Classical Mechanics. 2nd ed.
61 WHITEHEAD. Elements of Homotopy Theory.
62 KARGAPOLOV/MERLZJAKOV. Fundamentals of the Theory of Groups.
63 BOLLOBAS. Graph Theory.
64 EDWARDS. Fourier Series. Vol. I. 2nd ed.
65 WELLS. Differential Analysis on Complex Manifolds. 2nd ed.

(continued after index)

Bruce E. Sagan

The Symmetric Group

Representations, Combinatorial
Algorithms, and Symmetric Functions

Second Edition

With 31 Figures

 Springer

Bruce E. Sagan
Department of Mathematics
Michigan State University
East Lansing, MI 48824-1027
USA

Mathematics Subject Classification (2000): 20Cxx, 20C30, 20C32, 05C85

Library of Congress Cataloging-in-Publication Data
Sagan, Bruce Eli.
   The symmetric group : representations, combinatorial algorithms, and symmetric
functions / Bruce E. Sagan.
     p. cm. — (Graduate texts in mathematics ; 203)
   Includes bibliographical references and index.
   ISBN 978-1-4419-2869-6
   1. Representations of groups.   2. Symmetric functions.   I. Title.   II. Series.
QA171 .S24   2000
512'.2—dc21   00-040042

Printed on acid-free paper.

Production managed by Timothy Taylor; manufacturing supervised by Jeffrey Taub.
Photocomposed copy prepared from the author's LaTeX files.

Printed in the United States of America.

9 8 7 6 5 4 3 2 1

Springer-Verlag   New York Berlin Heidelberg
A member of BertelsmannSpringer Science+Business Media GmbH

Preface to the 2nd Edition

I have been very gratified by the response to the first edition, which has resulted in it being sold out. This put some pressure on me to come out with a second edition and now, finally, here it is.

The original text has stayed much the same, the major change being in the treatment of the hook formula which is now based on the beautiful Novelli-Pak-Stoyanovskii bijection [NPS 97]. I have also added a chapter on applications of the material from the first edition. This includes Stanley's theory of differential posets [Stn 88, Stn 90] and Fomin's related concept of growths [Fom 86, Fom 94, Fom 95], which extends some of the combinatorics of S_n-representations. Next come a couple of sections showing how groups acting on posets give rise to interesting representations that can be used to prove unimodality results [Stn 82]. Finally, we discuss Stanley's symmetric function analogue of the chromatic polynomial of a graph [Stn 95, Stn ta].

I would like to thank all the people, too numerous to mention, who pointed out typos in the first edition. My computer has been severely reprimanded for making them. Thanks also go to Christian Krattenthaler, Tom Roby, and Richard Stanley, all of whom read portions of the new material and gave me their comments. Finally, I would like to give my heartfelt thanks to my editor at Springer, Ina Lindemann, who has been very supportive and helpful through various difficult times.

Ann Arbor, Michigan, 2000

Preface to the 1st Edition

In recent years there has been a resurgence of interest in representations of symmetric groups (as well as other Coxeter groups). This topic can be approached from three directions: by applying results from the general theory of group representations, by employing combinatorial techniques, or by using symmetric functions. The fact that this area is the confluence of several strains of mathematics makes it an exciting one in which to study and work. By the same token, it is more difficult to master.

The purpose of this monograph is to bring together, for the first time under one cover, many of the important results in this field. To make the work accessible to the widest possible audience, a minimal amount of prior knowledge is assumed. The only prerequisites are a familiarity with elementary group theory and linear algebra. All other results about representations, combinatorics, and symmetric functions are developed as they are needed. Hence this book could be read by a graduate student or even a very bright undergraduate. For researchers I have also included topics from recent journal articles and even material that has not yet been published.

Chapter 1 is an introduction to group representations, with special emphasis on the methods of use when working with the symmetric groups. Because of space limitations, many important topics that are not germane to the rest of the development are not covered. These subjects can be found in any of the standard texts on representation theory.

In Chapter 2, the results from the previous chapter are applied to the symmetric group itself, and more highly specialized machinery is developed to handle this case. I have chosen to take the elegant approach afforded by the Specht modules rather than working with idempotents in the group algebra.

The third chapter focuses on combinatorics. It starts with the two famous formulae for the dimensions of the Specht modules: the Frame-Robinson-Thrall hook formula and the Frobenius-Young determinantal formula. The centerpiece is the Robinson-Schensted-Knuth algorithm, which allows us to describe some of the earlier theorems in purely combinatorial terms. A thorough discussion of Schützenberger's jeu de taquin and related matters is included.

Chapter 4 recasts much of the previous work in the language of symmetric functions. Schur functions are introduced, first combinatorially as the generating functions for semistandard tableaux and then in terms of symmetric group characters. The chapter concludes with the famous Littlewood-Richardson and Murnaghan-Nakayama rules.

My debt to several other books will be evident. Much of Chapter 1 is based on Ledermann's exquisite text on group characters [Led 77]. Chapter

2 borrows heavily from the monograph of James [Jam 78], whereas Chapter 4 is inspired by Macdonald's already classic book [Mac 79]. Finally, the third chapter is a synthesis of material from the research literature.

There are numerous applications of representations of groups, and in particular of the symmetric group, to other areas. For example, they arise in physics [Boe 70], probability and statistics [Dia 88], topological graph theory [Whi 84], and the theory of partially ordered sets [Stn 82]. However, to keep the length of this text reasonable, I have discussed only the connections with combinatorial algorithms.

This book grew out of a course that I taught while visiting the Université du Québec à Montréal during the fall of 1986. I would like to thank *l'équipe de combinatoire* for arranging my stay. I also presented this material in a class here at Michigan State University in the winter and spring of 1990. I thank my students in both courses for many helpful suggestions (and those at UQAM for tolerating my bad French). Francesco Brenti, Kathy Dempsey, Yoav Dvir, Kathy Jankoviak, and Scott Mathison have all pointed out ways in which the presentation could be improved. I also wish to express my appreciation of John Kimmel, Marlene Thom, and Linda Loba at Wadsworth and Brooks/Cole for their help during the preparation of the manuscript. Because I typeset this document myself, all errors can be blamed on my computer.

East Lansing, Michigan, 1991

Contents

Preface to the 2nd Edition   v

Preface to the 1st Eition   vii

1 Group Representations   **1**
   1.1 Fundamental Concepts   1
   1.2 Matrix Representations   4
   1.3 G-Modules and the Group Algebra   6
   1.4 Reducibility 10
   1.5 Complete Reducibility and Maschke's
       Theorem 13
   1.6 G-Homomorphisms and Schur's Lemma 18
   1.7 Commutant and Endomorphism Algebras 23
   1.8 Group Characters 30
   1.9 Inner Products of Characters 33
   1.10 Decomposition of the Group Algebra 40
   1.11 Tensor Products Again 43
   1.12 Restricted and Induced Representations 45
   1.13 Exercises 48

2 Representations of the Symmetric Group   **53**
   2.1 Young Subgroups, Tableaux, and Tabloids 53
   2.2 Dominance and Lexicographic Ordering 57
   2.3 Specht Modules 60
   2.4 The Submodule Theorem 63
   2.5 Standard Tableaux and a Basis for S^λ 66
   2.6 Garnir Elements 70
   2.7 Young's Natural Representation 74
   2.8 The Branching Rule 76
   2.9 The Decomposition of M^μ 78
   2.10 The Semistandard Basis for $\mathrm{Hom}(S^\lambda, M^\mu)$ 82
   2.11 Kostka Numbers and Young's Rule 85
   2.12 Exercises 86

3 Combinatorial Algorithms   **91**
   3.1 The Robinson-Schensted Algorithm 91
   3.2 Column Insertion 95
   3.3 Increasing and Decreasing Subsequences 97
   3.4 The Knuth Relations 99
   3.5 Subsequences Again 102
   3.6 Viennot's Geometric Construction 106
   3.7 Schützenberger's Jeu de Taquin 112
   3.8 Dual Equivalence 116
   3.9 Evacuation 121
   3.10 The Hook Formula 124
   3.11 The Determinantal Formula 132
   3.12 Exercises 133

4 Symmetric Functions   **141**
   4.1 Introduction to Generating Functions 142
   4.2 The Hillman-Grassl Algorithm 147
   4.3 The Ring of Symmetric Functions 151
   4.4 Schur Functions 155
   4.5 The Jacobi-Trudi Determinants 158
   4.6 Other Definitions of the Schur Function 163
   4.7 The Characteristic Map 167
   4.8 Knuth's Algorithm 169
   4.9 The Littlewood-Richardson Rule 174
   4.10 The Murnaghan-Nakayama Rule 179
   4.11 Exercises 185

5 Applications and Generalizations   **191**
   5.1 Young's Lattice and Differential Posets 191
   5.2 Growths and Local Rules 197
   5.3 Groups Acting on Posets 204
   5.4 Unimodality 208
   5.5 Chromatic Symmetric Functions 213
   5.6 Exercises 218

Bibliography   **223**

Index   **230**

List of Symbols

Symbol	Meaning	Page
A	poset	58
A_k	kth rank of a graded poset	195
A_S	rank selected poset	204
a_μ	alternant corresponding to composition μ	164
$\alpha \backslash \alpha_1$	composition α with first part α_1 removed	180
β	set partition	215
$\mathbb{C}$	complex numbers	4
ch^n	characteristic map	167
C_n	cyclic group of order n	5
C_t	column-stabilizer group of Young tableau t	60
$\mathrm{Com}\, X$	commutant algebra of representation X	23
$\mathbb{C}[\mathbf{G}]$	group algebra of G over $\mathbb{C}$	8
$\mathbb{C}[\mathbf{S}]$	vector space generated by set S over $\mathbb{C}$	7
$\mathbb{C}[[x]]$	ring of formal power series in x over $\mathbb{C}$	142
c_x	column insertion operator for integer x	95
$c_{\mu\nu}^\lambda$	Littlewood-Richardson coefficient	175
$\chi(g)$	character of group element g	30
χ_K	character value on conjugacy class K	32
χ^{def}	defining character of S_n	31
χ^{reg}	regular character	31
$\chi \downarrow_H^G$	restriction of character χ from G to H	45
$\chi \uparrow_H^G$	induction of character χ from h to G	45
D	down operator of a poset	196
$d_k(\pi)$	length of π's longest k-decreasing subsequence	103
ΔQ	delta operator applied to tableau Q	121
$\deg X$	degree of representation X	4
$\mathrm{Des}\, P$	the descent set of a tableau P	206
e_λ	λth elementary symmetric function	154
e_n	nth elementary symmetric function	152
e_t	polytabloid associated with Young tableau t	61
$E(\Gamma)$	edges of graph Γ	213

Symbol	Meaning	Page
$\text{End}\,V$	endomorphism algebra of module V	23
$\text{ev}\,Q$	the evacuation tableau of Q	122
ϵ	identity element of a group	1
f_S	weight generating function of weighted set S	145
f^λ	number of standard λ-tableaux	73
ϕ^λ	character of permutation module M^λ	56
G	group	1
$g_{A,B}$	Garnir element of a pair of sets A, B	70
GL_d	$d \times d$ complex general linear group	4
GP	generalized permutations	169
GP'	generalized permutations, no repeated columns	172
Γ	graph	213
h_n	nth complete homogeneous symmetric function	152
h_λ	λth complete homogeneous symmetric function	154
$\text{Hom}(V, W)$	all module homomorphisms from V to W	22
$h_{i,j}$	hooklength of cell (i, j)	124
$H_{i,j}$	hook of cell (i, j)	124
h_v	hooklength of cell v	124
H_v	hook of cell v	124
I	identity matrix	23
I_d	$d \times d$ identity matrix	24
$i_k(\pi)$	length of π's longest k-increasing subsequence	103
$i_\lambda(V)$	number of stable partitions of V of type λ	215
$\text{im}\,\theta$	image of θ	21
$j(P)$	jeu de taquin tableau of P	116
$j_c(P)$	forward slide on tableau P into cell c	113
$j^c(P)$	backward slide on tableau P into cell c	113
$j_Q(P)$	sequence of forward slides on P into Q	118
$j^P(Q)$	sequence of backward slides on Q into P	118
$\ker\theta$	kernel of θ	21
K_g	conjugacy class of a group element g	3
k_λ	size of K_λ	3
K_λ	conjugacy class in S_n corresponding to λ	3
$K_{\lambda\mu}$	Kostka number	85
κ_t	signed column sum for Young tableau t	61

Symbol	Meaning	Page
$l(\lambda)$	length (number of parts) of partition λ	152
$ll(\xi)$	leg length of rim hook ξ	180
λ	integer partition	2
	integer composition	67
λ'	conjugate or transpose of partition λ	103
$(\lambda_1, \lambda_2, \ldots, \lambda_l)$	partition given as a sequence	2
	composition given as a sequence	15
λ/μ	skew shape	112
$\lambda\backslash\xi$	partition λ with rim hook ξ removed	180
$\lambda(\beta)$	type of the set partition β	215
Λ	algebra of symmetric functions	151
Λ_l	symmetric functions in l variables	163
Λ^n	symmetric functions homogeneous of degree n	152
Mat_d	full $d \times d$ matrix algebra	4
m_λ	λth monomial symmetric function	151
M^λ	permutation module corresponding to λ	56
P	partial tableau	92
$p(n)$	number of partitions of n	143
p_n	nth power sum symmetric function	152
p_λ	λth power sum symmetric function	154
$P(\pi)$	P-tableau of permutation π	93
P_Γ	chromatic polynomial of graph Γ	214
$\mathbb{P}$	positive integers	214
π	permutation	1
π_P	row word of the Young tableau P	101
π^r	reversal of permutation π	97
$\hat{\pi}$	top row of generalized permutation π	169
$\check{\pi}$	bottom row of generalized permutation π	169
$\pi \stackrel{\text{R-S}}{\longleftrightarrow} (P,Q)$	Robinson-Schensted map	92
$\pi \stackrel{\text{R-S-K}}{\longleftrightarrow} (P,Q)$	Robinson-Schensted-Knuth map	170
$\pi \stackrel{\text{R-S-K}'}{\longleftrightarrow} (P,Q)$	dual Robinson-Schensted-Knuth map	172
$Q(\pi)$	Q-tableau of permutation π	93
$R(G)$	vector space of class functions on G	32
R^n	class functions of $\mathcal{S}_n$	167
R_t	row-stabilizer group of Young tableau t	60
r_x	row-insertion operator for the integer x	93
rk	the rank function in a poset	195

Symbol	Meaning	Page
S_A	group of permutations of A	20
s_λ	Schur function associated with λ	155
$s_{\lambda/\mu}$	skew Schur function associated with λ/μ	175
S_λ	Young subgroup associated with partition λ	54
S^λ	Specht module associated with partition λ	62
$\mathrm{sgn}(\pi)$	sign of permutation π	4
$\mathrm{sh}\, t$	shape of Young tableau t	55
S_n	symmetric group on $\{1,2,\ldots,n\}$	1
t	Young tableau	55
T	generalized Young tableau	78
$\{t\}$	row tabloid of t	55
$[t]$	column tabloid of t	72
$\mathcal{T}_{\lambda\mu}$	generalized tableaux, shape λ, content μ	78
$\mathcal{T}^0_{\lambda\mu}$	semistandard tableaux, shape λ, content μ	81
θ	homomorphism of modules	18
θ_T	homomorphism corresponding to tableau T	80
$\bar{\theta}_T$	restriction of θ_T to a Specht module	81
U	up operator of a poset	196
$v_Q(P)$	vacating tableau for $j_Q(P)$	118
$v^P(Q)$	vacating tableau for $j^P(Q)$	118
$V(\Gamma)$	vertices of graph Γ	213
$W^\perp$	orthogonal complement of W	15
wt	weight function	145
$\mathbf{x}$	the set of variables $\{x_1, x_2, x_3, \ldots\}$	151
$X(g)$	matrix of g in the representation X	4
$\mathbf{x}^T$	monomial weight of a tableau T	155
$\mathbf{x}^\mu$	monomial weight of a composition μ	155
$X\!\downarrow^G_H$	restriction of representation X from G to H	45
$X\!\uparrow^G_H$	induction of representation X from H to G	45
X_Γ	chromatic symmetric function of graph Γ	214
ξ	rim or skew hook	180
Y	Young's lattice	192
Z_A	center of algebra A	27
Z_g	centralizer of group element g	3
z_λ	size of Z_g where $g \in S_n$ has cycle type λ	3
$\mathbb{Z}$	integers	204

Symbol	Meaning	Page
$\hat{0}$	minimum of a poset	195
$\hat{1}$	maximum of a poset	204
1_G	trivial representation of G	5
$(1^{m_1}, 2^{m_2}, \ldots)$	partition given by multiplicities	2
$\cong$	equivalence of modules	19
	slide equivalence of tableaux	114
$\overset{*}{\cong}$	dual equivalence of tableaux	117
$\overset{K}{\cong}$	Knuth equivalence	100
$\overset{K^*}{\cong}$	dual Knuth equivalence	111
$\overset{P}{\cong}$	P-equivalence	99
$\overset{Q}{\cong}$	Q-equivalence	111
$\overset{1}{\cong}$	Knuth relation of the first kind	99
$\overset{1^*}{\cong}$	dual Knuth relation of the first kind	111
$\overset{2}{\cong}$	Knuth relation of the second kind	99
$\overset{2^*}{\cong}$	dual Knuth relation of the second kind	111
$\leq$	subgroup relation	9
	submodule relation	10
	lexicographic order on partitions	59
$\prec$	covering relation in a poset	58
$\trianglerighteq$	dominance order on partitions	58
	dominance order on tabloids	68
$\vdash$	is a partition of for integers	53
	is a partition of for sets	215
$\oplus$	direct sum of matrices	13
	direct sum of vector spaces	13
$\otimes$	tensor product of matrices	25
	tensor product of representations	43
	tensor product of vector spaces	26
$\langle \chi, \psi \rangle$	inner product of characters χ and ψ	34
$\chi \cdot \psi$	product of characters	168
$A \times B$	product of posets	198
$G \wr H$	wreath product of groups	212
$\cup$	union of tableaux	120
$\uplus$	disjoint union	45
$\lvert \cdot \rvert$	cardinality of a set	3
	sum of parts of a partition	54
$\wedge$	meet operation in a poset	192
$\vee$	join operation in a poset	192

Chapter 1

Group Representations

We begin our study of the symmetric group by considering its representations. First, however, we must present some general results about group representations that will be useful in our special case. Representation theory can be couched in terms of matrices or in the language of modules. We consider both approaches and then turn to the associated theory of characters. All our work will use the complex numbers as the ground field in order to make life as easy as possible.

We are presenting the material in this chapter so that this book will be relatively self-contained, although it all can be found in other standard texts. In particular, our exposition is modeled on the one in Ledermann [Led 77].

1.1 Fundamental Concepts

In this section we introduce some basic terminology and notation. We pay particular attention to the symmetric group.

Let G be a group written multiplicatively with identity ϵ. Throughout this work, G is finite unless stated otherwise. We assume that the reader is familiar with the elementary properties of groups (cosets, Lagrange's theorem, etc.) that can be found in any standard text such as Herstein [Her 64].

Our object of study is the *symmetric group*, S_n, consisting of all bijections from $\{1, 2, \ldots, n\}$ to itself using composition as the multiplication. The elements $\pi \in S_n$ are called *permutations*. We multiply permutations from right to left. (In fact, we compose all functions in this manner.) Thus $\pi\sigma$ is the bijection obtained by first applying σ, followed by π.

If π is a permutation, then there are three different notations we can use for this element. *Two-line notation* is the array

$$\pi = \begin{matrix} 1 & 2 & \cdots & n \\ \pi(1) & \pi(2) & \cdots & \pi(n) \end{matrix}.$$

1

For example, if $\pi \in S_5$ is given by

$$\pi(1) = 2, \qquad \pi(2) = 3, \qquad \pi(3) = 1, \qquad \pi(4) = 4, \qquad \pi(5) = 5,$$

then its two-line form is

$$\pi = \begin{matrix} 1 & 2 & 3 & 4 & 5 \\ 2 & 3 & 1 & 4 & 5 \end{matrix}.$$

Because the top line is fixed, we can drop it to get *one-line notation*.

Lastly, we can display π using *cycle notation*. Given $i \in \{1, 2, \ldots, n\}$, the elements of the sequence $i, \pi(i), \pi^2(i), \pi^3(i), \ldots$ cannot all be distinct. Taking the first power p such that $\pi^p(i) = i$, we have the cycle

$$(i, \pi(i), \pi^2(i), \ldots, \pi^{p-1}(i)).$$

Equivalently, the cycle $(i, j, k, \ldots, l)$ means that π sends i to j, j to k, $\ldots$, and l back to i. Now pick an element not in the cycle containing i and iterate this process until all members of $\{1, 2, \ldots, n\}$ have been used. Our example from the last paragraph becomes

$$\pi = (1, 2, 3)(4)(5)$$

in cycle notation. Note that cyclically permuting the elements within a cycle or reordering the cycles themselves does not change the permutation. Thus

$$(1, 2, 3)(4)(5) = (2, 3, 1)(4)(5) = (4)(2, 3, 1)(5) = (4)(5)(3, 1, 2).$$

A *k-cycle*, or *cycle of length k*, is a cycle containing k elements. The preceding permutation consists of a 3-cycle and two 1-cycles. The *cycle type*, or simply the *type*, of π is an expression of the form

$$(1^{m_1}, 2^{m_2}, \ldots, n^{m_n}),$$

where m_k is the number of cycles of length k in π. The example permutation has cycle type

$$(1^2, 2^0, 3^1, 4^0, 5^0).$$

A 1-cycle of π is called a *fixedpoint*. The numbers 4 and 5 are fixedpoints in our example. Fixedpoints are usually dropped from the cycle notation if no confusion will result. An *involution* is a permutation such that $\pi^2 = \epsilon$. It is easy to see that π is an involution if and only if all of π's cycles have length 1 or 2.

Another way to give the cycle type is as a partition. A *partition of n* is a sequence

$$\lambda = (\lambda_1, \lambda_2, \ldots, \lambda_l)$$

where the λ_i are weakly decreasing and $\sum_{i=1}^{l} \lambda_i = n$. Thus k is repeated m_k times in the partition version of the cycle type of π. Our example corresponds to the partition

$$\lambda = (3, 1, 1).$$

In any group G, elements g and h are *conjugates* if

$$g = khk^{-1}$$

for some $k \in G$. The set of all elements conjugate to a given g is called the *conjugacy class of g* and is denoted by K_g. Conjugacy is an equivalence relation, so the distinct conjugacy classes partition G. (This is a *set* partition, as opposed to the *integer* partitions discussed in the previous paragraph.) Returning to $\mathcal{S}_n$, it is not hard to see that if

$$\pi = (i_1, i_2, \ldots, i_l) \cdots (i_m, i_{m+1}, \ldots, i_n)$$

in cycle notation, then for any $\sigma \in \mathcal{S}_n$

$$\sigma \pi \sigma^{-1} = (\sigma(i_1), \sigma(i_2), \ldots, \sigma(i_l)) \cdots (\sigma(i_m), \sigma(i_{m+1}), \ldots, \sigma(i_n)).$$

It follows that two permutations are in the same conjugacy class if and only if they have the same cycle type. Thus there is a natural one-to-one correspondence between partitions of n and conjugacy classes of $\mathcal{S}_n$.

We can compute the size of a conjugacy class in the following manner. Let G be any group and consider the *centralizer* of $g \in G$ defined by

$$Z_g = \{ h \in G \ : \ hgh^{-1} = g \},$$

i.e., the set of all elements that commute with g. Now, there is a bijection between the cosets of Z_g and the elements of K_g, so that

$$|K_g| = \frac{|G|}{|Z_g|}, \tag{1.1}$$

where $| \cdot |$ denotes cardinality. Now let $G = \mathcal{S}_n$ and use K_λ for K_g when g has type λ.

Proposition 1.1.1 *If $\lambda = (1^{m_1}, 2^{m_2}, \ldots, n^{m_n})$ and $g \in \mathcal{S}_n$ has type λ, then $|Z_g|$ depends only on λ and*

$$z_\lambda \stackrel{\text{def}}{=} |Z_g| = 1^{m_1} m_1! 2^{m_2} m_2! \cdots n^{m_n} m_n!$$

Proof. Any $h \in Z_g$ can either permute the cycles of length i among themselves or perform a cyclic rotation on each of the individual cycles (or both). Since there are $m_i!$ ways to do the former operation and i^{m_i} ways to do the latter, we are done. ∎

Thus equation (1.1) specializes in the symmetric group to

$$k_\lambda = \frac{n!}{z_\lambda} = \frac{n!}{1^{m_1} m_1! 2^{m_2} m_2! \cdots n^{m_n} m_n!}, \tag{1.2}$$

where $k_\lambda = |K_\lambda|$.

Of particular interest is the conjugacy class of *transpositions*, which are those permutations of the form $\tau = (i, j)$. The transpositions generate $\mathcal{S}_n$ as a group; in fact, the symmetric group is generated by the *adjacent transpositions* $(1, 2)$, $(2, 3)$, ..., $(n - 1, n)$. If $\pi = \tau_1 \tau_2 \cdots \tau_k$, where the τ_i are transpositions, then we define the *sign of* π to be

$$\text{sgn}(\pi) = (-1)^k.$$

It can be proved that sgn is well defined, i.e., independent of the particular decomposition of π into transpositions. Once this is established, it follows easily that

$$\text{sgn}(\pi\sigma) = \text{sgn}(\pi)\,\text{sgn}(\sigma). \tag{1.3}$$

As we will see, this is an example of a representation.

1.2   Matrix Representations

A matrix representation can be thought of as a way to model an abstract group with a concrete group of matrices. After giving the precise definition, we look at some examples.

Let $\mathbb{C}$ denote the complex numbers. Let Mat_d stand for the set of all $d \times d$ matrices with entries in $\mathbb{C}$. This is called a *full complex matrix algebra of degree d*. Recall that an algebra is a vector space with an associative multiplication of vectors (thus also imposing a ring structure on the space). The *complex general linear group of degree d*, denoted by GL_d, is the group of all $X = (x_{i,j})_{d \times d} \in \text{Mat}_d$ that are invertible with respect to multiplication.

Definition 1.2.1 A *matrix representation of a group G* is a group homomorphism

$$X : G \to GL_d.$$

Equivalently, to each $g \in G$ is assigned $X(g) \in \text{Mat}_d$ such that

1. $X(\epsilon) = I$ the identity matrix, and

2. $X(gh) = X(g)X(h)$ for all $g, h \in G$.

The parameter d is called the *degree*, or *dimension*, of the representation and is denoted by $\deg X$. ∎

Note that conditions 1 and 2 imply that $X(g^{-1}) = X(g)^{-1}$, so these matrices are in GL_d as required.

Obviously the simplest representations are those of degree 1. Our first two examples are of this type.

Example 1.2.2 All groups have the *trivial representation*, which is the one sending every $g \in G$ to the matrix (1). This is clearly a representation because $X(\epsilon) = (1)$ and

$$X(g)X(h) = (1)(1) = (1) = X(gh)$$

for all $g, h \in G$. We often use 1_G or just the number 1 itself to stand for the trivial representation of G. ∎

Example 1.2.3 Let us find all one-dimensional representations of the cyclic group of order n, C_n. Let g be a generator of C_n, i.e.,

$$C_n = \{g, g^2, g^3, \ldots, g^n = \epsilon\}.$$

If $X(g) = (c)$, $c \in \mathbb{C}$, then the matrix for every element of C_n is determined, since $X(g^k) = (c^k)$ by property 2 in the preceding definition. But by property 1,

$$(c^n) = X(g^n) = X(\epsilon) = (1),$$

so c must be an nth root of unity. Clearly, each such root gives a representation, so there are exactly n representations of C_n having degree 1.

In particular, let $n = 4$ and $C_4 = \{\epsilon, g, g^2, g^3\}$. The four fourth roots of unity are $1, i, -1, -i$. If we let the four corresponding representations be denoted by $X^{(1)}, X^{(2)}, X^{(3)}, X^{(4)}$, then we can construct a table:

	ϵ	g	g^2	g^3
$X^{(1)}$	1	1	1	1
$X^{(2)}$	1	i	-1	$-i$
$X^{(3)}$	1	-1	1	-1
$X^{(4)}$	1	$-i$	-1	i

where the entry in row i and column j is $X^{(i)}(g^j)$ (matrix brackets omitted). This array is an example of a character table, a concept we will develop in Section 1.8. (For representations of dimension 1, the representation and its character coincide.) Note that the trivial representation forms the first row of the table.

There are other representations of C_4 of larger degree. For example, we can let

$$X(g) = \begin{pmatrix} 1 & 0 \\ 0 & i \end{pmatrix}.$$

However, this representation is really just a combination of $X^{(1)}$ and $X^{(2)}$. In the language of Section 1.5, X is completely reducible with irreducible components $X^{(1)}$ and $X^{(2)}$. We will see that every representation of C_n can be constructed in this way using the n representations of degree 1 as building blocks. ∎

Example 1.2.4 We have already met a nontrivial degree 1 representation of S_n. In fact, equation (1.3) merely says that the map $X(\pi) = (\text{sgn}(\pi))$ is a representation, called the *sign representation*.

Also of importance is the *defining representation* of S_n, which is of degree n. If $\pi \in S_n$, then we let $X(\pi) = (x_{i,j})_{n \times n}$, where

$$x_{i,j} = \begin{cases} 1 & \text{if } \pi(j) = i, \\ 0 & \text{otherwise.} \end{cases}$$

The matrix $X(\pi)$ is called a *permutation matrix*, since it contains only zeros and ones, with a unique one in every row and column. The reader should verify that this is a representation.

In particular, consider $\mathcal{S}_3$ with its permutations written in cycle notation. Then the matrices of the defining representation are

$$X(\epsilon) = \begin{pmatrix} 1 & 0 & 0 \\ 0 & 1 & 0 \\ 0 & 0 & 1 \end{pmatrix}, \quad X((1,2)) = \begin{pmatrix} 0 & 1 & 0 \\ 1 & 0 & 0 \\ 0 & 0 & 1 \end{pmatrix},$$

$$X((1,3)) = \begin{pmatrix} 0 & 0 & 1 \\ 0 & 1 & 0 \\ 1 & 0 & 0 \end{pmatrix}, \quad X((2,3)) = \begin{pmatrix} 1 & 0 & 0 \\ 0 & 0 & 1 \\ 0 & 1 & 0 \end{pmatrix},$$

$$X((1,2,3)) = \begin{pmatrix} 0 & 0 & 1 \\ 1 & 0 & 0 \\ 0 & 1 & 0 \end{pmatrix}, \quad X((1,3,2)) = \begin{pmatrix} 0 & 1 & 0 \\ 0 & 0 & 1 \\ 1 & 0 & 0 \end{pmatrix}. \quad \blacksquare$$

1.3 G-Modules and the Group Algebra

Because matrices correspond to linear transformations, we can think of representations in these terms. This is the idea of a G-module.

Let V be a vector space. Unless stated otherwise, all vector spaces will be over the complex numbers and of finite dimension. Let $GL(V)$ stand for the set of all invertible linear transformations of V to itself, called the *general linear group of V*. If $\dim V = d$, then $GL(V)$ and GL_d are isomorphic as groups.

Definition 1.3.1 Let V be a vector space and G be a group. Then V is a *G-module* if there is a group homomorphism

$$\rho : G \to GL(V).$$

Equivalently, V is a G-module if there is a multiplication, $g\mathbf{v}$, of elements of V by elements of G such that

1. $g\mathbf{v} \in V$,

2. $g(c\mathbf{v} + d\mathbf{w}) = c(g\mathbf{v}) + d(g\mathbf{w})$,

3. $(gh)\mathbf{v} = g(h\mathbf{v})$, and

4. $\epsilon\mathbf{v} = \mathbf{v}$

for all $g, h \in G$; $\mathbf{v}, \mathbf{w} \in V$; and scalars $c, d \in \mathbb{C}$. $\blacksquare$

In the future, "G-module" will often be shortened to "module" when no confusion can result about the group involved. Other words involving G- as a prefix will be treated similarly. Alternatively, we can say that the vector space V *carries* a representation of G.

Let us verify that the two parts of the definition are equivalent. In fact, we are just using $g\mathbf{v}$ as a shorthand for the application of the transformation $\rho(g)$ to the vector $\mathbf{v}$. Item 1 says that the transformation takes V to itself; 2 shows that the map is linear; 3 is property 2 of the matrix definition; and 4 in combination with 3 says that g and g^{-1} are inverse maps, so all transformations are invertible. Although it is more abstract than our original definition of representation, the G-module concept lends itself to cleaner proofs.

In fact, we can go back and forth between these two notions of representation quite easily. Given a matrix representation X of degree d, let V be the vector space $\mathbb{C}^d$ of all column vectors of length d. Then we can multiply $\mathbf{v} \in V$ by $g \in G$ using the definition

$$g\mathbf{v} \overset{\text{def}}{=} X(g)\mathbf{v},$$

where the operation on the right is matrix multiplication. Conversely, if V is a G-module, then take any basis $\mathcal{B}$ for V. Thus $X(g)$ will just be the matrix of the linear transformation $g \in G$ in the basis $\mathcal{B}$ computed in the usual way. We use this correspondence extensively in the rest of this book.

Group actions are important in their own right. Note that if S is any set with a multiplication by elements of G satisfying 1, 3, and 4, then we say G *acts on* S. In fact, it is always possible to take a set on which G acts and turn it into a G-module as follows. Let $S = \{s_1, s_2, \ldots, s_n\}$ and let $\mathbb{C}S = \mathbb{C}\{s_1, s_2, \ldots, s_n\}$ denote the vector space generated by S over $\mathbb{C}$; i.e., $\mathbb{C}S$ consists of all the formal linear combinations

$$c_1\mathbf{s}_1 + c_2\mathbf{s}_2 + \cdots + c_n\mathbf{s}_n,$$

where $c_i \in \mathbb{C}$ for all i. (We put the elements of S in boldface print when they are being considered as vectors.) Vector addition and scalar multiplication in $\mathbb{C}S$ are defined by

$$(c_1\mathbf{s}_1 + c_2\mathbf{s}_2 + \cdots + c_n\mathbf{s}_n) + (d_1\mathbf{s}_1 + d_2\mathbf{s}_2 + \cdots + d_n\mathbf{s}_n)$$
$$= (c_1 + d_1)\mathbf{s}_1 + (c_2 + d_2)\mathbf{s}_2 + \cdots + (c_n + d_n)\mathbf{s}_n$$

and

$$c\,(c_1\mathbf{s}_1 + c_2\mathbf{s}_2 + \cdots + c_n\mathbf{s}_n) = (cc_1)\mathbf{s}_1 + (cc_2)\mathbf{s}_2 + \cdots + (cc_n)\mathbf{s}_n,$$

respectively. Now the action of G on S can be extended to an action on $\mathbb{C}S$ by linearity:

$$g\,(c_1\mathbf{s}_1 + c_2\mathbf{s}_2 + \cdots + c_n\mathbf{s}_n) = c_1(g\mathbf{s}_1) + c_2(g\mathbf{s}_2) + \cdots + c_n(g\mathbf{s}_n)$$

for all $g \in G$. This makes $\mathbb{C}S$ into a G-module of dimension $|S|$.

Definition 1.3.2 If a group G acts on a set S, then the associated module $\mathbb{C}S$ is called the *permutation representation* associated with S. Also, the elements of S form a basis for $\mathbb{C}S$ called the *standard basis*. ∎

All the following examples of G-modules are of this form.

Example 1.3.3 Consider the symmetric group $\mathcal{S}_n$ with its usual action on $S = \{1, 2, \ldots, n\}$. Now

$$\mathbb{C}S = \{c_1 \mathbf{1} + c_2 \mathbf{2} + \cdots + c_n \mathbf{n} : c_i \in \mathbb{C} \text{ for all } i\}$$

with the action

$$\pi(c_1 \mathbf{1} + c_2 \mathbf{2} + \cdots + c_n \mathbf{n}) = c_1 \pi(\mathbf{1}) + c_2 \pi(\mathbf{2}) + \cdots + c_n \pi(\mathbf{n})$$

for all $\pi \in \mathcal{S}_n$.

To make things more concrete, we can select a basis and determine the matrices $X(\pi)$ for $\pi \in \mathcal{S}_n$ in that basis. Let us consider $\mathcal{S}_3$ and use the standard basis $\{\mathbf{1}, \mathbf{2}, \mathbf{3}\}$. To find the matrix for $\pi = (1,2)$, we compute

$$(1,2)\mathbf{1} = \mathbf{2}; \qquad (1,2)\mathbf{2} = \mathbf{1}; \qquad (1,2)\mathbf{3} = \mathbf{3};$$

and so

$$X(\,(1,2)\,) = \begin{pmatrix} 0 & 1 & 0 \\ 1 & 0 & 0 \\ 0 & 0 & 1 \end{pmatrix}.$$

If the reader determines the rest of the matrices for $\mathcal{S}_3$, it will be noted that they are exactly the same as those of the defining representation, Example 1.2.4. It is not hard to show that the same is true for any n; i.e., this is merely the module approach to the defining representation of $\mathcal{S}_n$. ∎

Example 1.3.4 We now describe one of the most important representations for any group, the *(left) regular representation*. Let G be an arbitrary group. Then G acts on itself by left multiplication: if $g \in G$ and $h \in S = G$, then the action of g on h, gh, is defined as the usual product in the group. Properties 1, 3, and 4 now follow, respectively, from the closure, associativity, and identity axioms for the group.

Thus if $G = \{g_1, g_2, \ldots, g_n\}$, then we have the corresponding G-module

$$\mathbb{C}[\mathbf{G}] = \{c_1 \mathbf{g_1} + c_2 \mathbf{g_2} + \cdots + c_n \mathbf{g_n} : c_i \in \mathbb{C} \text{ for all } i\},$$

which is called the *group algebra of G*. Note the use of square brackets to indicate that this is an algebra, not just a vector space. The multiplication is gotten by letting $\mathbf{g_i}\mathbf{g_j} = \mathbf{g_k}$ in $\mathbb{C}[\mathbf{G}]$ if $g_i g_j = g_k$ in G, and linear extension. Now the action of G on the group algebra can be expressed as

$$g\,(c_1 \mathbf{g_1} + c_2 \mathbf{g_2} + \cdots + c_n \mathbf{g_n}) = c_1(\mathbf{g}\mathbf{g_1}) + c_2(\mathbf{g}\mathbf{g_2}) + \cdots + c_n(\mathbf{g}\mathbf{g_n})$$

for all $g \in G$. The group algebra will furnish us with much important combinatorial information about group representations.

Let us see what the regular representation of the cyclic group C_4 looks like. First of all,

$$\mathbb{C}[C_4] = \{c_1\epsilon + c_2\mathbf{g} + c_3\mathbf{g}^2 + c_4\mathbf{g}^3 \; : \; c_i \in \mathbb{C} \text{ for all } i\}.$$

We can easily find the matrix of g^2 in the standard basis:

$$g^2\epsilon = \mathbf{g}^2, \qquad g^2\mathbf{g} = \mathbf{g}^3, \qquad g^2\mathbf{g}^2 = \epsilon, \qquad g^2\mathbf{g}^3 = \mathbf{g}.$$

Thus

$$X(g^2) = \begin{pmatrix} 0 & 0 & 1 & 0 \\ 0 & 0 & 0 & 1 \\ 1 & 0 & 0 & 0 \\ 0 & 1 & 0 & 0 \end{pmatrix}.$$

Computing the rest of the matrices would be a good exercise. Note that they are all permutation matrices and all distinct. In general, the regular representation of G gives an embedding of G into the symmetric group on $|G|$ elements. The reader has probably seen this presented in a group theory course, in a slightly different guise, as Cayley's theorem [Her 64, pages 60–61].

Note that if G acts on any V, then so does $\mathbb{C}[G]$. Specifically, if $c_1\mathbf{g}_1 + c_2\mathbf{g}_2 + \cdots + c_n\mathbf{g}_n \in \mathbb{C}[G]$ and $\mathbf{v} \in V$, then we can define the action

$$(c_1\mathbf{g}_1 + c_2\mathbf{g}_2 + \cdots + c_n\mathbf{g}_n)\mathbf{v} = c_1(g_1\mathbf{v}) + c_2(g_2\mathbf{v}) + \cdots + c_n(g_n\mathbf{v}).$$

In fact, we can extend the concept of representation to algebras: A representation of an algebra A is an algebra homomorphism from A into $GL(V)$. In this way, every representation of a group G gives rise to a representation of its group algebra $\mathbb{C}[G]$. For a further discussion of representations of algebras, see the text of Curtis and Reiner [C-R 66]. ∎

Example 1.3.5 Let group G have subgroup H, written $H \leq G$. A generalization of the regular representation is the *(left) coset representation* of G with respect to H. Let $g_1, g_2, \ldots, g_k$ be a transversal for H; i.e., $\mathcal{H} = \{g_1H, g_2H, \ldots, g_kH\}$ is a complete set of disjoint left cosets for H in G. Then G acts on $\mathcal{H}$ by letting

$$g(g_iH) = (gg_i)H$$

for all $g \in G$. The corresponding module

$$\mathbb{C}\mathcal{H} = \{c_1\mathbf{g}_1\mathbf{H} + c_2\mathbf{g}_2\mathbf{H} + \cdots + c_k\mathbf{g}_k\mathbf{H} \; : \; c_i \in \mathbb{C} \text{ for all } i\}$$

inherits the action

$$g(c_1\mathbf{g}_1\mathbf{H} + \cdots + c_k\mathbf{g}_k\mathbf{H}) = c_1(\mathbf{g}\mathbf{g}_1\mathbf{H}) + \cdots + c_k(\mathbf{g}\mathbf{g}_k\mathbf{H}).$$

Note that if $H = G$, then this reduces to the trivial representation. At the other extreme, when $H = \{\epsilon\}$, then $\mathcal{H} = G$ and we obtain the regular representation again. In general, representation by cosets is an example of an induced representation, an idea studied further in Section 1.12.

Let us consider $G = S_3$ and $H = \{\epsilon, (2,3)\}$. We can take

$$\mathcal{H} = \{H, \ (1,2)H, \ (1,3)H\}$$

and

$$\mathbb{C}\mathcal{H} = \{c_1 H + c_2(1,2)H + c_3(1,3)H \ : \ c_i \in \mathbb{C} \text{ for all } i\}.$$

Computing the matrix of $(1,2)$ in the standard basis, we obtain

$$(1,2)\mathbf{H} = (1,2)\mathbf{H}, \ (1,2)(1,2)\mathbf{H} = \mathbf{H}, \ (1,2)(1,3)\mathbf{H} = (1,3,2)\mathbf{H} = (1,3)\mathbf{H},$$

so that

$$X(\ (1,2)\ ) = \begin{pmatrix} 0 & 1 & 0 \\ 1 & 0 & 0 \\ 0 & 0 & 1 \end{pmatrix}.$$

After finding a few more matrices, the reader will become convinced that we have rediscovered the defining representation yet again. The reason for this is explained when we consider isomorphism of modules in Section 1.6. ∎

1.4   Reducibility

An idea pervading all of science is that large structures can be understood by breaking them up into their smallest pieces. The same thing is true of representation theory. Some representations are built out of smaller ones (such as the one at the end of Example 1.2.3), whereas others are indivisible (as are all degree one representations). This is the distinction between reducible and irreducible representations, which we study in this section. First, however, we must determine precisely what a piece or subobject means in this setting.

Definition 1.4.1 Let V be a G-module. A *submodule of V* is a subspace W that is closed under the action of G, i.e.,

$$\mathbf{w} \in W \Rightarrow g\mathbf{w} \in W \text{ for all } g \in G.$$

We also say that W is a *G-invariant subspace*. Equivalently, W is a subset of V that is a G-module in its own right. We write $W \leq V$ if W is a submodule of V. ∎

As usual, we illustrate the definition with some examples.

Example 1.4.2 Any G-module, V, has the submodules $W = V$ as well as $W = \{\mathbf{0}\}$, where $\mathbf{0}$ is the zero vector. These two submodules are called *trivial*. All other submodules are called *nontrivial*. ∎

Example 1.4.3 For a nontrivial example of a submodule, consider $G = \mathcal{S}_n$, $n \geq 2$, and $V = \mathbb{C}\{\mathbf{1}, \mathbf{2}, \ldots, \mathbf{n}\}$ (the defining representation). Now take

$$W = \mathbb{C}\{\mathbf{1} + \mathbf{2} + \cdots + \mathbf{n}\} = \{c(\mathbf{1} + \mathbf{2} + \cdots + \mathbf{n}) \ : \ c \in \mathbb{C}\};$$

i.e., W is the one-dimensional subspace spanned by the vector $\mathbf{1} + \mathbf{2} + \cdots + \mathbf{n}$. To check that W is closed under the action of $\mathcal{S}_n$, it suffices to show that

$$\pi \mathbf{w} \in W \text{ for all } \mathbf{w} \text{ in some basis for } W \text{ and all } \pi \in \mathcal{S}_n.$$

(Why?) Thus we need to verify only that

$$\pi(\mathbf{1} + \mathbf{2} + \cdots + \mathbf{n}) \in W$$

for each $\pi \in \mathcal{S}_n$. But

$$\pi(\mathbf{1} + \mathbf{2} + \cdots + \mathbf{n}) = \pi(\mathbf{1}) + \pi(\mathbf{2}) + \cdots + \pi(\mathbf{n})$$
$$= \mathbf{1} + \mathbf{2} + \cdots + \mathbf{n} \in W,$$

because applying π to $\{1, 2, \ldots, n\}$ just gives back the same set of numbers in a different order. Thus W is a submodule of V that is nontrivial since $\dim W = 1$ and $\dim V = n \geq 2$.

Since W is a module for G sitting inside V, we can ask what representation we get if we restrict the action of G to W. But we have just shown that every $\pi \in \mathcal{S}_n$ sends the basis vector $\mathbf{1} + \mathbf{2} + \cdots + \mathbf{n}$ to itself. Thus $X(\pi) = (1)$ is the corresponding matrix, and we have found a copy of the trivial representation in $\mathbb{C}\{\mathbf{1}, \mathbf{2}, \ldots, \mathbf{n}\}$. In general, for a vector space W of any dimension, if G fixes every element of W, we say that G *acts trivially on* W. ∎

Example 1.4.4 Next, let us look again at the regular representation. Suppose $G = \{g_1, g_2, \ldots, g_n\}$ with group algebra $V = \mathbb{C}[\mathbf{G}]$. Using the same idea as in the previous example, let

$$W = \mathbb{C}[\mathbf{g}_1 + \mathbf{g}_2 + \cdots + \mathbf{g}_n],$$

the one-dimensional subspace spanned by the vector that is the sum of all the elements of G. To verify that W is a submodule, take any $g \in G$ and compute:

$$g(\mathbf{g}_1 + \mathbf{g}_2 + \cdots + \mathbf{g}_n) = g\mathbf{g}_1 + g\mathbf{g}_2 + \cdots + g\mathbf{g}_n$$
$$= \mathbf{g}_1 + \mathbf{g}_2 + \cdots + \mathbf{g}_n \in W,$$

because multiplying by g merely permutes the elements of G, leaving the sum unchanged. As before, G acts trivially on W.

The reader should verify that if $G = \mathcal{S}_n$, then the sign representation can also be recovered by using the submodule

$$W = \mathbb{C}[\textstyle\sum_{\pi \in \mathcal{S}_n} \operatorname{sgn}(\pi) \ \pi \ ] \quad \blacksquare$$

We now introduce the irreducible representations that will be the building blocks of all the others.

Definition 1.4.5 A nonzero G-module V is *reducible* if it contains a nontrivial submodule W. Otherwise, V is said to be *irreducible*. Equivalently, V is reducible if it has a basis $\mathcal{B}$ in which every $g \in G$ is assigned a block matrix of the form

$$X(g) = \left(\begin{array}{c|c} A(g) & B(g) \\ \hline 0 & C(g) \end{array} \right) \qquad (1.4)$$

where the $A(g)$ are square matrices, all of the same size, and 0 is a nonempty matrix of zeros. ∎

To see the equivalence, suppose V of dimension d has a submodule W of dimension f, $0 < f < d$. Then let

$$\mathcal{B} = \{\mathbf{w}_1, \mathbf{w}_2, \ldots, \mathbf{w}_f, \mathbf{v}_{f+1}, \mathbf{v}_{f+2}, \ldots, \mathbf{v}_d\},$$

where the first f vectors are a basis for W. Now we can compute the matrix of any $g \in G$ with respect to the basis $\mathcal{B}$. Since W is a submodule, $g\mathbf{w}_i \in W$ for all i, $1 \leq i \leq f$. Thus the last $d - f$ coordinates of $g\mathbf{w}_i$ will all be zero. That accounts for the zero matrix in the lower left corner of $X(g)$. Note that we have also shown that the $A(g)$, $g \in G$, are the matrices of the restriction of G to W. Hence they must all be square and of the same size.

Conversely, suppose each $X(g)$ has the given form with every $A(g)$ being $f \times f$. Let $V = \mathbb{C}^d$ and consider

$$W = \mathbb{C}\{\mathbf{e}_1, \mathbf{e}_2, \ldots, \mathbf{e}_f\},$$

where $\mathbf{e}_i$ is the column vector with a 1 in the ith row and zeros elsewhere (the *standard basis* for $\mathbb{C}^d$). Then the placement of the zeros in $X(g)$ assures us that $X(g)\mathbf{e}_i \in W$ for $1 \leq i \leq f$ and all $g \in G$. Thus W is a G-module, and it is nontrivial because the matrix of zeros is nonempty.

Clearly, any epresentation of degree 1 is irreducible. It seems hard to determine when a representation of greater degree will be irreducible. Certainly, checking all possible subspaces to find out which ones are submodules is out of the question. This unsatisfactory state of affairs will be remedied after we discuss inner products of group characters in Section 1.9.

From the preceding examples, both the defining representation for $\mathcal{S}_n$ and the group algebra for an arbitrary G are reducible if $n \geq 2$ and $|G| \geq 2$, respectively. After all, we produced nontrivial submodules. Let us now illustrate the alternative approach via matrices using the defining representation of $\mathcal{S}_3$. We must extend the basis $\{1 + 2 + 3\}$ for W to a basis for $V = \mathbb{C}\{1, 2, 3\}$. Let us pick

$$\mathcal{B} = \{1 + 2 + 3, \mathbf{2}, \mathbf{3}\}.$$

Of course, $X(e)$ remains the 3×3 identity matrix. To compute $X(\,(1,2)\,)$, we look at $(1,2)$'s action on our basis:

$$(1,2)(1 + 2 + 3) = 1 + 2 + 3, \ (1,2)2 = 1 = (1 + 2 + 3) - 2 - 3, \ (1,2)3 = 3.$$

So

$$X(\,(1,2)\,) = \begin{pmatrix} 1 & 1 & 0 \\ 0 & -1 & 0 \\ 0 & -1 & 1 \end{pmatrix}.$$

The reader can do the similar computations for the remaining four elements of S_3 to verify that

$$X(\,(1,3)\,) = \begin{pmatrix} 1 & 0 & 1 \\ 0 & 1 & -1 \\ 0 & 0 & -1 \end{pmatrix}, \qquad X(\,(2,3)\,) = \begin{pmatrix} 1 & 0 & 0 \\ 0 & 0 & 1 \\ 0 & 1 & 0 \end{pmatrix},$$

$$X(\,(1,2,3)\,) = \begin{pmatrix} 1 & 0 & 1 \\ 0 & 0 & -1 \\ 0 & 1 & -1 \end{pmatrix}, \qquad X(\,(1,3,2)\,) = \begin{pmatrix} 1 & 1 & 0 \\ 0 & -1 & 1 \\ 0 & -1 & 0 \end{pmatrix}.$$

Note that all these matrices have the form

$$X(\pi) = \left(\begin{array}{c|cc} 1 & * & * \\ \hline 0 & * & * \\ 0 & * & * \end{array} \right).$$

The one in the upper left corner comes from the fact that S_3 acts trivially on W.

1.5 Complete Reducibility and Maschke's Theorem

It would be even better if we could bring the matrices of a reducible G-module to the block diagonal form

$$X(g) = \left(\begin{array}{c|c} A(g) & 0 \\ \hline 0 & B(g) \end{array} \right)$$

for all $g \in G$. This is the notion of a direct sum.

Definition 1.5.1 Let V be a vector space with subspaces U and W. Then V is the *(internal) direct sum of U and W*, written $V = U \oplus W$, if every $\mathbf{v} \in V$ can be written uniquely as a sum

$$\mathbf{v} = \mathbf{u} + \mathbf{w}, \qquad \mathbf{u} \in U, \ \mathbf{w} \in W.$$

If V is a G-module and U, W are G-submodules, then we say that U *and* W *are complements of each other.*

If X is a matrix, then X is the *direct sum of matrices A and B*, written $X = A \oplus B$, if X has the block diagonal form

$$X = \left(\begin{array}{c|c} A & 0 \\ \hline 0 & B \end{array} \right). \ \blacksquare$$

To see the relationship between the module and matrix definitions, let V be a G-module with $V = U \oplus W$, where $U, W \le V$. Since this is a direct sum of vector spaces, we can choose a basis for V

$$\mathcal{B} = \{\mathbf{u}_1, \mathbf{u}_2, \ldots, \mathbf{u}_f, \mathbf{w}_{f+1}, \mathbf{w}_{f+2}, \ldots, \mathbf{w}_d\}$$

such that $\{\mathbf{u}_1, \mathbf{u}_2, \ldots, \mathbf{u}_f\}$ is a basis for U and $\{\mathbf{w}_{f+1}, \mathbf{w}_{f+2}, \ldots, \mathbf{w}_d\}$ is a basis for W. Since U and W are submodules, we have

$$g\mathbf{u}_i \in U \quad \text{and} \quad g\mathbf{w}_j \in W$$

for all $g \in G$, $\mathbf{u}_i \in U$, $\mathbf{w}_j \in W$. Thus the matrix of any $g \in G$ in the basis $\mathcal{B}$ is

$$X(g) = \left(\begin{array}{c|c} A(g) & 0 \\ \hline 0 & B(g) \end{array} \right),$$

where $A(g)$ and $B(g)$ are the matrices of the action of G restricted to U and W, respectively.

Returning to the defining representation of $\mathcal{S}_3$, we see that

$$V = \mathbb{C}\{1, 2, 3\} = \mathbb{C}\{1 + 2 + 3\} \oplus \mathbb{C}\{2, 3\}$$

as vector spaces. But while $\mathbb{C}\{1 + 2 + 3\}$ is an $\mathcal{S}_3$-submodule, $\mathbb{C}\{2, 3\}$ is not (e.g., $(1, 2)2 = 1 \notin \mathbb{C}\{2, 3\}$). So we need to find a complement for $\mathbb{C}\{1 + 2 + 3\}$, i.e., a *submodule* U such that

$$\mathbb{C}\{1, 2, 3\} = \mathbb{C}\{1 + 2 + 3\} \oplus U.$$

To find a complement, we introduce an inner product on $\mathbb{C}\{1, 2, 3\}$. Given any two vectors $\mathbf{i}, \mathbf{j}$ in the basis $\{1, 2, 3\}$, let their inner product be

$$\langle \mathbf{i}, \mathbf{j} \rangle = \delta_{i,j}, \tag{1.5}$$

where $\delta_{i,j}$ is the Kronecker delta. Now we extend by linearity in the first variable and conjugate linearity in the second to obtain an inner product on the whole vector space. Equivalently, we could have started out by defining the product of any two given vectors $\mathbf{v} = a1 + b2 + c3$, $\mathbf{w} = x1 + y2 + z3$ as

$$\langle \mathbf{v}, \mathbf{w} \rangle = a\bar{x} + b\bar{y} + c\bar{z},$$

with the bar denoting complex conjugation. The reader can check that this definition does indeed satisfy all the axioms for an inner product. It also enjoys the property that it is *invariant* under the action of G:

$$\langle g\mathbf{v}, g\mathbf{w} \rangle = \langle \mathbf{v}, \mathbf{w} \rangle \qquad \text{for all } g \in G \text{ and } \mathbf{v}, \mathbf{w} \in V. \tag{1.6}$$

To check invariance on V, it suffices to verify (1.6) for elements of a basis. But if $\pi \in \mathcal{S}_3$, then

$$\langle \pi\mathbf{i}, \pi\mathbf{j} \rangle = \delta_{\pi(i), \pi(j)} = \delta_{i,j} = \langle \mathbf{i}, \mathbf{j} \rangle,$$

where the middle equality holds because π is a bijection.

Now, given any inner product on a vector space V and a subspace W, we can form the *orthogonal complement*:

$$W^\perp = \{\mathbf{v} \in V \ : \ \langle \mathbf{v}, \mathbf{w} \rangle = 0 \text{ for all } \mathbf{w} \in W\}.$$

It is always true that $V = W \oplus W^\perp$. When $W \leq V$ and the inner product is G-invariant, we can say more.

Proposition 1.5.2 *Let V be a G-module, W a submodule, and $\langle \cdot, \cdot \rangle$ an inner product invariant under the action of G. Then $W^\perp$ is also a G-submodule.*

Proof. We must show that for all $g \in G$ and $\mathbf{u} \in W^\perp$ we have $g\mathbf{u} \in W^\perp$. Take any $\mathbf{w} \in W$; then

$$
\begin{aligned}
\langle g\mathbf{u}, \mathbf{w} \rangle &= \langle g^{-1}g\mathbf{u}, g^{-1}\mathbf{w} \rangle && \text{(since } \langle \cdot, \cdot \rangle \text{ is invariant)} \\
&= \langle \mathbf{u}, g^{-1}\mathbf{w} \rangle && \text{(properties of group action)} \\
&= 0. && (\mathbf{u} \in W^\perp, \text{ and } g^{-1}\mathbf{w} \in W \\
& && \text{since } W \text{ is a submodule)}
\end{aligned}
$$

Thus $W^\perp$ is closed under the action of G. ∎

Applying this to our running example, we see that

$$
\begin{aligned}
\mathbb{C}\{\mathbf{1} + \mathbf{2} + \mathbf{3}\}^\perp &= \{\mathbf{v} = a\mathbf{1} + b\mathbf{2} + c\mathbf{3} \ : \ \langle \mathbf{v}, \mathbf{1} + \mathbf{2} + \mathbf{3} \rangle = 0\} \\
&= \{\mathbf{v} = a\mathbf{1} + b\mathbf{2} + c\mathbf{3} \ : \ a + b + c = 0\}.
\end{aligned}
$$

To compute the matrices of the direct sum, we choose the bases $\{\mathbf{1} + \mathbf{2} + \mathbf{3}\}$ for $\mathbb{C}\{\mathbf{1} + \mathbf{2} + \mathbf{3}\}$, and $\{\mathbf{2} - \mathbf{1}, \ \mathbf{3} - \mathbf{1}\}$ for $\mathbb{C}\{\mathbf{1} + \mathbf{2} + \mathbf{3}\}^\perp$. This produces the matrices

$$
X(e) = \begin{pmatrix} 1 & 0 & 0 \\ 0 & 1 & 0 \\ 0 & 0 & 1 \end{pmatrix}, \qquad X((1,2)) = \begin{pmatrix} 1 & 0 & 0 \\ 0 & -1 & -1 \\ 0 & 0 & 1 \end{pmatrix},
$$

$$
X((1,3)) = \begin{pmatrix} 1 & 0 & 0 \\ 0 & 1 & 0 \\ 0 & -1 & -1 \end{pmatrix}, \qquad X((2,3)) = \begin{pmatrix} 1 & 0 & 0 \\ 0 & 0 & 1 \\ 0 & 1 & 0 \end{pmatrix},
$$

$$
X((1,2,3)) = \begin{pmatrix} 1 & 0 & 0 \\ 0 & -1 & -1 \\ 0 & 1 & 0 \end{pmatrix}, \qquad X((1,3,2)) = \begin{pmatrix} 1 & 0 & 0 \\ 0 & 0 & 1 \\ 0 & -1 & -1 \end{pmatrix}.
$$

These are indeed all direct matrix sums of the form

$$
X(g) = \left(\begin{array}{c|cc} A(g) & 0 & 0 \\ \hline 0 & & \\ 0 & & B(g) \end{array} \right).
$$

Of course, $A(g)$ is irreducible (being of degree 1), and we will see in Section 1.9 that $B(g)$ is also. Thus we have decomposed the defining representation of S_3 into its irreducible parts. The content of Maschke's theorem is that this can be done for any finite group.

Theorem 1.5.3 (Maschke's Theorem) *Let G be a finite group and let V be a nonzero G-module. Then*

$$V = W^{(1)} \oplus W^{(2)} \oplus \cdots \oplus W^{(k)},$$

where each $W^{(i)}$ is an irreducible G-submodule of V.

Proof. We will induct on $d = \dim V$. If $d = 1$, then V itself is irreducible and we are done ($k = 1$ and $W^{(1)} = V$). Now suppose that $d > 1$. If V is irreducible, then we are finished as before. If not, then V has a nontrivial G-submodule, W. We will try to construct a submodule complement for W as we did in the preceding example.

Pick any basis $\mathcal{B} = \{\mathbf{v}_1, \mathbf{v}_2, \ldots, \mathbf{v}_d\}$ for V. Consider the unique inner product that satisfies

$$\langle \mathbf{v}_i, \mathbf{v}_j \rangle = \delta_{i,j}$$

for elements of $\mathcal{B}$. This product may not be G-invariant, but we can come up with another one that is. For any $\mathbf{v}, \mathbf{w} \in V$ we let

$$\langle \mathbf{v}, \mathbf{w} \rangle' = \sum_{g \in G} \langle g\mathbf{v}, g\mathbf{w} \rangle.$$

We leave it to the reader to verify that $\langle \cdot, \cdot \rangle'$ satisfies the definition of an inner product. To show that it is G-invariant, we wish to prove

$$\langle h\mathbf{v}, h\mathbf{w} \rangle' = \langle \mathbf{v}, \mathbf{w} \rangle'$$

for all $h \in G$ and $\mathbf{v}, \mathbf{w} \in V$. But

$$
\begin{aligned}
\langle h\mathbf{v}, h\mathbf{w} \rangle' &= \textstyle\sum_{g \in G} \langle gh\mathbf{v}, gh\mathbf{w} \rangle && \text{(definition of } \langle \cdot, \cdot \rangle') \\
&= \textstyle\sum_{f \in G} \langle f\mathbf{v}, f\mathbf{w} \rangle && \text{(as } g \text{ varies over } G, \text{ so does } f = gh) \\
&= \langle \mathbf{v}, \mathbf{w} \rangle' && \text{(definition of } \langle \cdot, \cdot \rangle')
\end{aligned}
$$

as desired.

If we let

$$W^{\perp} = \{\mathbf{v} \in V \ : \ \langle \mathbf{v}, \mathbf{w} \rangle' = 0\},$$

then by Proposition 1.5.2 we have that $W^{\perp}$ is a G-submodule of V with

$$V = W \oplus W^{\perp}.$$

Now we can apply induction to W and $W^{\perp}$ to write each as a direct sum of irreducibles. Putting these two decompositions together, we see that V has the desired form. ∎

As a corollary, we have the matrix version of Maschke's theorem. Here and in the future, we often drop the horizontal and vertical lines indicating block matrices. Our convention of using lowercase letters for elements and uppercase ones for matrices should avoid any confusion.

Corollary 1.5.4 *Let G be a finite group and let X be a matrix representation of G of dimension $d > 0$. Then there is a fixed matrix T such that every matrix $X(g)$, $g \in G$, has the form*

$$TX(g)T^{-1} = \begin{pmatrix} X^{(1)}(g) & 0 & \cdots & 0 \\ 0 & X^{(2)}(g) & \cdots & 0 \\ \vdots & \vdots & \ddots & \vdots \\ 0 & 0 & \cdots & X^{(k)}(g) \end{pmatrix},$$

where each $X^{(i)}$ is an irreducible matrix representation of G.

Proof. Let $V = \mathbb{C}^d$ with the action

$$g\mathbf{v} = X(g)\mathbf{v}$$

for all $g \in G$ and $\mathbf{v} \in V$. By Maschke's theorem,

$$V = W^{(1)} \oplus W^{(2)} \oplus \cdots \oplus W^{(k)},$$

each $W^{(i)}$ being irreducible of dimension, say, d_i. Take a basis $\mathcal{B}$ for V such that the first d_1 vectors are a basis for $W^{(1)}$, the next d_2 are a basis for $W^{(2)}$, etc. The matrix T that transforms the standard basis for $\mathbb{C}^d$ into $\mathcal{B}$ now does the trick, since conjugating by T just expresses each $X(g)$ in the new basis $\mathcal{B}$. ∎

Representations that decompose so nicely have a name.

Definition 1.5.5 A representation is *completely reducible* if it can be written as a direct sum of irreducibles. ∎

So Maschke's theorem could be restated:

> *Every representation of a finite group having positive dimension is completely reducible.*

We are working under the nicest possible assumptions, namely, that all our groups are finite and all our vector spaces are over $\mathbb{C}$. We will, however, occasionally attempt to indicate more general results. Maschke's theorem remains true if $\mathbb{C}$ is replaced by any field whose characteristic is either zero or prime to $|G|$. For a proof in this setting, the reader can consult Ledermann [Led 77, pages 21–23].

However, we can *not* drop the finiteness assumption on G, as the following example shows. Let $\mathbf{R}^+$ be the positive real numbers, which are a group under multiplication. It is not hard to see that letting

$$X(r) = \begin{pmatrix} 1 & \log\ r \\ 0 & 1 \end{pmatrix}$$

for all $r \in \mathbf{R}^+$ defines a representation. The subspace

$$W = \left\{ \begin{pmatrix} c \\ 0 \end{pmatrix} : c \in \mathbb{C} \right\} \subset \mathbb{C}^2$$

is invariant under the action of G. Thus if X is completely reducible, then $\mathbb{C}^2$ must decompose as the direct sum of W and another one-dimensional submodule. By the matrix version of Maschke's theorem, there exists a fixed matrix T such that

$$TX(r)T^{-1} = \begin{pmatrix} x(r) & 0 \\ 0 & y(r) \end{pmatrix}$$

for all $r \in \mathbf{R}^+$. Thus $x(r)$ and $y(r)$ must be the eigenvalues of $X(r)$, which are both 1. But then

$$X(r) = T^{-1} \begin{pmatrix} 1 & 0 \\ 0 & 1 \end{pmatrix} T = \begin{pmatrix} 1 & 0 \\ 0 & 1 \end{pmatrix}$$

for all $r \in \mathbf{R}^+$, which is absurd.

1.6 G-Homomorphisms and Schur's Lemma

We can learn more about objects in mathematics (e.g., vector spaces, groups, topological spaces) by studying functions that preserve their structure (e.g., linear transformations, homomorphisms, continuous maps). For a G-module, the corresponding function is called a G-homomorphism.

Definition 1.6.1 Let V and W be G-modules. Then a *G-homomorphism* (or simply a *homomorphism*) is a linear transformation $\theta : V \rightarrow W$ such that

$$\theta(g\mathbf{v}) = g\theta(\mathbf{v})$$

for all $g \in G$ and $\mathbf{v} \in V$. We also say that θ *preserves* or *respects* the action of G. ∎

We can translate this into the language of matrices by taking bases $\mathcal{B}$ and $\mathcal{C}$ for V and W, respectively. Let $X(g)$ and $Y(g)$ be the corresponding matrix representations. Also, take T to be the matrix of θ in the two bases $\mathcal{B}$ and $\mathcal{C}$. Then the G-homomorphism property becomes

$$TX(g)\mathbf{v} = Y(g)T\mathbf{v}$$

for every column vector $\mathbf{v}$ and $g \in G$. But since this holds for all $\mathbf{v}$, we must have

$$TX(g) = Y(g)T \qquad \text{for all } g \in G. \tag{1.7}$$

Thus having a G-homomorphism θ is equivalent to the existence of a matrix T such that (1.7) holds. We will often write this condition simply as $TX = YT$.

As an example, let $G = \mathcal{S}_n$, $V = \mathbb{C}\{\mathbf{v}\}$ with the trivial action of $\mathcal{S}_n$, and let $W = \mathbb{C}\{\mathbf{1}, \mathbf{2}, \ldots, \mathbf{n}\}$ with the defining action of $\mathcal{S}_n$. Define a transformation $\theta : V \rightarrow W$ by

$$\theta(\mathbf{v}) = \mathbf{1} + \mathbf{2} + \cdots + \mathbf{n}$$

and linear extension; i.e.,

$$\theta(c\mathbf{v}) = c(\mathbf{1} + \mathbf{2} + \cdots + \mathbf{n})$$

for all $c \in \mathbb{C}$. To check that θ is a G-homomorphism, it suffices to check that the action of G is preserved on a basis of V. (Why?) But for all $\pi \in \mathcal{S}_n$,

$$\theta(\pi\mathbf{v}) = \theta(\mathbf{v}) = \sum_{i=1}^{n}\mathbf{i} = \pi\sum_{i=1}^{n}\mathbf{i} = \pi\theta(\mathbf{v}).$$

In a similar vein, let G be an arbitrary group acting trivially on $V = \mathbb{C}\{\mathbf{v}\}$, and let $W = \mathbb{C}[\mathbf{G}]$ be the group algebra. Now we have the G-homomorphism $\theta : V \to W$ given by extending

$$\theta(\mathbf{v}) = \sum_{g \in G}\mathbf{g}$$

linearly.

If $G = \mathcal{S}_n$, we can also let G act on $V = \mathbb{C}\{\mathbf{v}\}$ by using the sign representation:

$$\pi\mathbf{u} = \operatorname{sgn}(\pi)\mathbf{u}$$

for all $\pi \in \mathcal{S}_n$ and $\mathbf{u} \in V$. Keeping the usual action on the group algebra, the reader can verify that

$$\eta(\mathbf{v}) = \sum_{\pi \in \mathcal{S}_n} \operatorname{sgn}(\pi)\boldsymbol{\pi}$$

extends to a G-homomorphism from V to W.

It is clearly important to know when two representations of a group are different and when they are not (even though there may be some cosmetic differences). For example, two matrix representations that differ only by a basis change are really the same. The concept of G-equivalence captures this idea.

Definition 1.6.2 Let V and W be modules for a group G. A *G-isomorphism* is a G-homomorphism $\theta : V \to W$ that is bijective. In this case we say that V and W are *G-isomorphic*, or *G-equivalent*, written $V \cong W$. Otherwise we say that V and W are *G-inequivalent*. ∎

As usual, we drop the G when the group is implied by context.

In matrix terms, θ being a bijection translates into the corresponding matrix T being invertible. Thus from equation (1.7) we see that matrix representations X and Y of a group G are equivalent if and only if there exists a fixed matrix T such that

$$Y(g) = TX(g)T^{-1}$$

for all $g \in G$. This is the change-of-basis criterion that we were talking about earlier.

Example 1.6.3 We are now in a position to explain why the coset representation of S_3 at the end of Example 1.3.5 is the same as the defining representation. Recall that we had taken the subgroup $H = \{\epsilon, (2,3)\} \subset S_3$ giving rise to the coset representation module $\mathbb{C}\mathcal{H}$, where

$$\mathcal{H} = \{H, (1,2)H, (1,3)H\}.$$

Given any set A, let S_A be the symmetric group on A, i.e., the set of all permutations of T. Now the subgroup H can be expressed as an (internal) direct product

$$H = \{(1)(2)(3), (1)(2,3)\} = \{(1)\} \times \{(2)(3), (2,3)\} = S_{\{1\}} \times S_{\{2,3\}}. \quad (1.8)$$

A convenient device for displaying such product subgroups of S_n is the tabloid. Let $\lambda = (\lambda_1, \lambda_2, \ldots, \lambda_l)$ be a partition, as discussed in Section 1.1. A *Young tabloid of shape* λ is an array with l rows such that row i contains λ_i integers and the order of entries in a row not matter. To show that each row can be shuffled arbitrarily, we put horizontal lines between the rows. For example, if $\lambda = (4, 2, 1)$, then some of the possible Young tabloids are

$$
\begin{array}{cccc}
\underline{3} & \underline{1} & \underline{4} & \underline{1} \\
\underline{5} & \underline{9} & & \\
\underline{2} & & &
\end{array}
\quad = \quad
\begin{array}{cccc}
\underline{3} & \underline{1} & \underline{1} & \underline{4} \\
\underline{9} & \underline{5} & & \\
\underline{2} & & &
\end{array}
\quad \neq \quad
\begin{array}{cccc}
\underline{9} & \underline{5} & \underline{3} & \underline{4} \\
\underline{2} & \underline{1} & & \\
\underline{1} & & &
\end{array} .
$$

Equation (1.8) says that H consists of all permutations in S_3 that permute the elements of the set $\{1\}$ among themselves (giving only the permutation (1)) and permute the elements of $\{2,3\}$ among themselves (giving $(2)(3)$ and $(2,3)$). This is modeled by the tabloid

$$
\begin{array}{cc}
\underline{2} & \underline{3} \\
\underline{1} &
\end{array} ,
$$

since the order of 2 and 3 is immaterial but 1 must remain fixed. The complete set of tabloids of shape $\lambda = (2,1)$ whose entries are exactly 1, 2, 3 is

$$
S = \left\{
\begin{array}{cc}
\underline{2} & \underline{3} \\
\underline{1} &
\end{array} ,
\begin{array}{cc}
\underline{1} & \underline{3} \\
\underline{2} &
\end{array} ,
\begin{array}{cc}
\underline{1} & \underline{2} \\
\underline{3} &
\end{array}
\right\} .
$$

Furthermore, there is an action of any $\pi \in S_3$ on S given by

$$
\pi
\begin{array}{cc}
\underline{i} & \underline{j} \\
\underline{k} &
\end{array}
=
\begin{array}{cc}
\underline{\pi(i)} & \underline{\pi(j)} \\
\underline{\pi(k)} &
\end{array} .
$$

Thus it makes sense to consider the map θ that sends

$$
H \xrightarrow{\theta}
\begin{array}{cc}
\underline{2} & \underline{3} \\
\underline{1} &
\end{array} ,
$$

$$
(1,2)H \xrightarrow{\theta} (1,2)
\begin{array}{cc}
\underline{2} & \underline{3} \\
\underline{1} &
\end{array}
=
\begin{array}{cc}
\underline{1} & \underline{3} \\
\underline{2} &
\end{array} ,
$$

$$
(1,3)H \xrightarrow{\theta} (1,3)
\begin{array}{cc}
\underline{2} & \underline{3} \\
\underline{1} &
\end{array}
=
\begin{array}{cc}
\underline{1} & \underline{2} \\
\underline{3} &
\end{array} .
$$

By linear extension, θ becomes a vector space isomorphism from $\mathbb{C}\mathcal{H}$ to $\mathbb{C}\mathbf{S}$. In fact, we claim it is also a G-isomorphism. To verify this, we can check that the action of each $\pi \in \mathcal{S}_3$ is preserved on each basis vector in H. For example, if $\pi = (1,2)$ and $\mathbf{H} \in \mathcal{H}$, then

$$\theta((1,2)\mathbf{H}) = \theta(\overline{(1,2)\mathbf{H}}) = \overline{\begin{array}{cc} 1 & 3 \\ 2 \end{array}} = (1,2)\overline{\begin{array}{cc} 2 & 3 \\ 1 \end{array}} = (1,2)\theta(\mathbf{H}).$$

Thus

$$\mathbb{C}\mathcal{H} \cong \mathbb{C}\mathbf{S}. \tag{1.9}$$

Another fact about the tabloids in our set S is that they are completely determined by the element placed in the second row. So we have a natural map, η, between the basis $\{\mathbf{1}, \mathbf{2}, \mathbf{3}\}$ for the defining representation and $\mathbf{S}$, namely,

$$\mathbf{1} \overset{\eta}{\to} \overline{\begin{array}{cc} 2 & 3 \\ 1 \end{array}},$$

$$\mathbf{2} \overset{\eta}{\to} \overline{\begin{array}{cc} 1 & 3 \\ 2 \end{array}},$$

$$\mathbf{3} \overset{\eta}{\to} \overline{\begin{array}{cc} 1 & 2 \\ 3 \end{array}}.$$

Now η extends by linearity to a G-isomorphism from $\mathbb{C}\{\mathbf{1}, \mathbf{2}, \mathbf{3}\}$ to $\mathbb{C}\mathbf{S}$. This, in combination with equation (1.9), shows that $\mathbb{C}\mathcal{H}$ and $\mathbb{C}\{\mathbf{1}, \mathbf{2}, \mathbf{3}\}$ are indeed equivalent.

The reader may feel that we have taken a long and winding route to get to the final $\mathcal{S}_3$-isomorphism. However, the use of Young tabloids extends far beyond this example. In fact, we use them to construct all the irreducible representations of $\mathcal{S}_n$ in Chapter 2. ∎

We now return to the general exposition. Two sets usually associated with any map of vector spaces $\theta : V \to W$ are its *kernel*,

$$\ker \theta = \{\mathbf{v} \in V \ : \ \theta(\mathbf{v}) = \mathbf{0}\},$$

where $\mathbf{0}$ is the zero vector, and its *image*,

$$\operatorname{im} \theta = \{\mathbf{w} \in W \ : \ \mathbf{w} = \theta(\mathbf{v}) \text{ for some } \mathbf{v} \in V\}.$$

When θ is a G-homomorphism, the kernel and image have nice structure.

Proposition 1.6.4 *Let $\theta : V \to W$ be a G-homomorphism. Then*

1. *$\ker \theta$ is a G-submodule of V, and*

2. *$\operatorname{im} \theta$ is a G-submodule of W.*

Proof. We prove only the first assertion, leaving the second one for the reader. It is known from the theory of vector spaces that $\ker \theta$ is a subspace of V since θ is linear. So we only need to show closure under the action of G. But if $\mathbf{v} \in \ker \theta$, then for any $g \in G$,

$$\begin{aligned} \theta(g\mathbf{v}) &= g\theta(\mathbf{v}) \quad (\theta \text{ is a } G\text{-homomorphism}) \\ &= g\mathbf{0} \quad\quad (\mathbf{v} \in \ker \theta) \\ &= \mathbf{0}, \end{aligned}$$

and so $g\mathbf{v} \in \ker \theta$, as desired. ∎

It is now an easy matter to prove Schur's lemma, which characterizes G-homomorphisms of irreducible modules. This result plays a crucial role when we discuss the commutant algebra in the next section.

Theorem 1.6.5 (Schur's Lemma) *Let V and W be two irreducible G-modules. If $\theta : V \to W$ is a G-homomorphism, then either*

1. *θ is a G-isomorphism, or*

2. *θ is the zero map.*

Proof. Since V is irreducible and $\ker \theta$ is a submodule (by the previous proposition), we must have either $\ker \theta = \{\mathbf{0}\}$ or $\ker \theta = V$. Similarly, the irreducibility of W implies that $\operatorname{im} \theta = \{\mathbf{0}\}$ or W. If $\ker \theta = V$ or $\operatorname{im} \theta = \{\mathbf{0}\}$, then θ must be the zero map. On the other hand, if $\ker \theta = \{\mathbf{0}\}$ and $\operatorname{im} \theta = W$, then we have an isomorphism. ∎

It is interesting to note that Schur's lemma continues to be valid over arbitrary fields and for infinite groups. In fact, the proof we just gave still works. The matrix version is also true in this more general setting.

Corollary 1.6.6 *Let X and Y be two irreducible matrix representations of G. If T is any matrix such that $TX(g) = Y(g)T$ for all $g \in G$, then either*

1. *T is invertible, or*

2. *T is the zero matrix.* ∎

We also have an analogue of Schur's lemma in the case where the range module is not irreducible. This result is conveniently expressed in terms of the vector space $\operatorname{Hom}(V, W)$ of all G-homomorphisms from V to W.

Corollary 1.6.7 *Let V and W be two G-modules with V being irreducible. Then $\dim \operatorname{Hom}(V, W) = 0$ if and only if W contains no submodule isomorphic to V.* ∎

When the field is $\mathbb{C}$, however, we can say more. Suppose that T is a matrix such that

$$TX(g) = X(g)T \tag{1.10}$$

for all $g \in G$. It follows that

$$(T - cI)X = X(T - cI),$$

where I is the appropriate identity matrix and $c \in \mathbb{C}$ is any scalar. Now $\mathbb{C}$ is algebraically closed, so we can take c to be an eigenvalue of T. Thus $T - cI$ satisfies the hypothesis of Corollary 1.6.6 (with $X = Y$) and is not invertible by the choice of c. Our only alternative is that $T - cI = 0$. We have proved the following result:

Corollary 1.6.8 *Let X be an irreducible matrix representation of G over the complex numbers. Then the only matrices T that commute with $X(g)$ for all $g \in G$ are those of the form $T = cI$—i.e., scalar multiples of the identity matrix.* ∎

1.7 Commutant and Endomorphism Algebras

Corollary 1.6.8 suggests that the set of matrices that commute with those of a given representation are important. This corresponds in the module setting to the set of G-homomorphisms from a G-module to itself. We characterize these sets in this section. Extending these ideas to homomorphisms between different G-modules leads to a useful generalization of Corollary 1.6.7 (see Corollary 1.7.10).

Definition 1.7.1 Given a matrix representation $X : G \rightarrow GL_d$, the corresponding *commutant algebra* is

$$\operatorname{Com} X = \{T \in \operatorname{Mat}_d \ : \ TX(g) = X(g)T \text{ for all } g \in G\},$$

where Mat_d is the set of all $d \times d$ matrices with entries in $\mathbb{C}$. Given a G-module V, the corresponding *endomorphism algebra* is

$$\operatorname{End} V = \{\theta : V \rightarrow V \ : \ \theta \text{ is a } G\text{-homomorphism}\}. \ \blacksquare$$

It is easy to check that both the commutant and endomorphism algebras do satisfy the axioms for an algebra. The reader can also verify that if V is a G-module and X is a corresponding matrix representation, then $\operatorname{End} V$ and $\operatorname{Com} X$ are isomorphic as algebras. Merely take the basis $\mathcal{B}$ that produced X and use the map that sends each $\theta \in \operatorname{End} V$ to the matrix T of θ in the basis $\mathcal{B}$. Let us compute $\operatorname{Com} X$ for various representations X.

Example 1.7.2 Suppose that X is a matrix representation such that

$$X = \left(\begin{array}{cc} X^{(1)} & 0 \\ 0 & X^{(2)} \end{array} \right) = X^{(1)} \oplus X^{(2)},$$

where $X^{(1)}, X^{(2)}$ are inequivalent and irreducible of degrees d_1, d_2, respectively. What does $\operatorname{Com} X$ look like?

Suppose that

$$T = \left(\begin{array}{cc} T_{1,1} & T_{1,2} \\ T_{2,1} & T_{2,2} \end{array} \right)$$

is a matrix partitioned in the same way as X. If $TX = XT$, then we can multiply out each side to obtain

$$\left(\begin{array}{cc} T_{1,1}X^{(1)} & T_{1,2}X^{(2)} \\ T_{2,1}X^{(1)} & T_{2,2}X^{(2)} \end{array} \right) = \left(\begin{array}{cc} X^{(1)}T_{1,1} & X^{(1)}T_{1,2} \\ X^{(2)}T_{2,1} & X^{(2)}T_{2,2} \end{array} \right).$$

Equating corresponding blocks we get

$$\begin{array}{rcl} T_{1,1}X^{(1)} & = & X^{(1)}T_{1,1}, \\ T_{1,2}X^{(2)} & = & X^{(1)}T_{1,2}, \\ T_{2,1}X^{(1)} & = & X^{(2)}T_{2,1}, \\ T_{2,2}X^{(2)} & = & X^{(2)}T_{2,2}. \end{array}$$

Using Corollaries 1.6.6 and 1.6.8 along with the fact that $X^{(1)}$ and $X^{(2)}$ are inequivalent, these equations can be solved to yield

$$T_{1,1} = c_1 I_{d_1}, \qquad T_{1,2} = T_{2,1} = 0, \qquad T_{2,2} = c_2 I_{d_2},$$

where $c_1, c_2 \in \mathbb{C}$ and I_{d_1}, I_{d_2} are identity matrices of degrees d_1, d_2. Thus

$$T = \left(\begin{array}{cc} c_1 I_{d_1} & 0 \\ 0 & c_2 I_{d_2} \end{array} \right).$$

We have shown that when $X = X^{(1)} \oplus X^{(2)}$ with $X^{(1)} \not\cong X^{(2)}$ and irreducible, then

$$\mathrm{Com}\, X = \{c_1 I_{d_1} \oplus c_2 I_{d_2} \ : \ c_1, c_2 \in \mathbb{C}\},$$

where $d_1 = \deg X^{(1)}$, $d_2 = \deg X^{(2)}$. $\blacksquare$

In general, if $X = \oplus_{i=1}^{k} X^{(i)}$, where the $X^{(i)}$ are pairwise inequivalent irreducibles, then a similar argument proves that

$$\mathrm{Com}\, X = \{\oplus_{i=1}^{k} c_i I_{d_i} \ : \ c_i \in \mathbb{C}\},$$

where $d_i = \deg X^{(i)}$. Notice that the degree of X is $\sum_{i=1}^{k} d_i$. Note also that the dimension of $\mathrm{Com}\, X$ (as a vector space) is just k. This is because there are k scalars c_i that can vary, whereas the identity matrices are fixed.

Next we deal with the case of sums of equivalent representations. A convenient notation is

$$mX = \overbrace{X \oplus X \oplus \cdots \oplus X}^{m},$$

where the nonnegative integer m is called the *multiplicity of* X.

Example 1.7.3 Suppose that

$$X = \begin{pmatrix} X^{(1)} & 0 \\ 0 & X^{(1)} \end{pmatrix} = 2X^{(1)},$$

where $X^{(1)}$ is irreducible of degree d. Take T partitioned as before. Doing the multiplication in $TX = XT$ and equating blocks now yields four equations, all of the form

$$T_{i,j}X^{(1)} = X^{(1)}T_{i,j}$$

for all $i, j = 1, 2$. Corollaries 1.6.6 and 1.6.8 come into play again to reveal that, for all i and j,

$$T_{i,j} = c_{i,j}I_d,$$

where $c_{i,j} \in \mathbb{C}$. Thus

$$\text{Com} \, X = \left\{ \begin{pmatrix} c_{1,1}I_d & c_{1,2}I_d \\ c_{2,1}I_d & c_{2,2}I_d \end{pmatrix} : c_{i,j} \in \mathbb{C} \text{ for all } i, j \right\} \qquad (1.11)$$

is the commutant algebra in this case. ∎

The matrices in $\text{Com} \, 2X^{(1)}$ have a name.

Definition 1.7.4 Let $X = (x_{i,j})$ and Y be matrices. Then their *tensor product* is the block matrix

$$X \otimes Y = (x_{i,j}Y) = \begin{pmatrix} x_{1,1}Y & x_{1,2}Y & \cdots \\ x_{2,1}Y & x_{2,2}Y & \cdots \\ \vdots & \vdots & \ddots \end{pmatrix}. \; \blacksquare$$

Thus we could write the elements of (1.11) as

$$T = \begin{pmatrix} c_{1,1} & c_{1,2} \\ c_{2,1} & c_{2,2} \end{pmatrix} \otimes I_d,$$

and so

$$\text{Com} \, X = \{M_2 \otimes I_d : M_2 \in \text{Mat}_2\}.$$

If we take $X = mX^{(1)}$, then

$$\text{Com} \, X = \{M_m \otimes I_d : M_m \in \text{Mat}_m\},$$

where d is the degree of $X^{(1)}$. Computing degrees and dimensions, we obtain

$$\deg X = \deg mX^{(1)} = m \deg X^{(1)} = md$$

and

$$\dim(\text{Com} \, X) = \dim\{M_m : M_m \in \text{Mat}_m\} = m^2.$$

Finally, we are led to consider the most general case:

$$X = m_1 X^{(1)} \oplus m_2 X^{(2)} \oplus \cdots \oplus m_k X^{(k)}, \qquad (1.12)$$

where the $X^{(i)}$ are pairwise inequivalent irreducibles with $\deg X^{(i)} = d_i$. The degree of X is given by

$$\deg X = \sum_{i=1}^{k} \deg(m_i X^{(i)}) = m_1 d_1 + m_2 d_2 + \cdots + m_k d_k.$$

The reader should have no trouble combining Examples 1.7.2 and 1.7.3 to obtain

$$\operatorname{Com} X = \{\oplus_{i=1}^{k}(M_{m_i} \otimes I_{d_i}) \; : \; M_{m_i} \in \operatorname{Mat}_{m_i} \text{ for all } i\} \tag{1.13}$$

of dimension

$$\dim(\operatorname{Com} X) = \dim\{\oplus_{i=1}^{k} M_{m_i} \; : \; M_{m_i} \in \operatorname{Mat}_{m_i}\} = m_1^2 + m_2^2 + \cdots + m_k^2.$$

Before continuing our investigation of the commutant algebra, we should briefly mention the abstract vector space analogue of the tensor product.

Definition 1.7.5 Given vector spaces V and W, then their *tensor product* is the set

$$V \otimes W = \{\sum_{i,j} c_{i,j} \mathbf{v}_i \otimes \mathbf{w}_j \; : \; c_{i,j} \in \mathbb{C}, \mathbf{v}_i \in V, \mathbf{w}_j \in W\}$$

subject to the relations

$$(c_1 \mathbf{v}_1 + c_2 \mathbf{v}_2) \otimes \mathbf{w} = c_1(\mathbf{v}_1 \otimes \mathbf{w}) + c_2(\mathbf{v}_2 \otimes \mathbf{w})$$

and

$$\mathbf{v} \otimes (d_1 \mathbf{w}_1 + d_2 \mathbf{w}_2) = d_1(\mathbf{v} \otimes \mathbf{w}_1) + d_2(\mathbf{v} \otimes \mathbf{w}_2). \ \blacksquare$$

It is easy to see that $V \otimes W$ is also a vector space. In fact, the reader can check that if $\mathcal{B} = \{\mathbf{v}_1, \mathbf{v}_2, \ldots, \mathbf{v}_d\}$ and $\mathcal{C} = \{\mathbf{w}_1, \mathbf{w}_2, \ldots, \mathbf{w}_f\}$ are bases for V and W, respectively, then the set

$$\{\mathbf{v}_i \otimes \mathbf{w}_j \; : \; 1 \leq i \leq d, 1 \leq j \leq f\}$$

is a basis for $V \otimes W$. This gives the connection with the definition of matrix tensor products: The algebra Mat_d has as basis the set

$$\mathcal{B} = \{E_{i,j} \; : \; 1 \leq i, j \leq d\},$$

where $E_{i,j}$ is the matrix of zeros with exactly one 1 in position (i, j). So if $X = (x_{i,j}) \in \operatorname{Mat}_d$ and $Y = (y_{k,l}) \in \operatorname{Mat}_f$, then, by the fact that $\otimes$ is linear,

$$\begin{aligned} X \otimes Y &= (\sum_{i,j=1}^{d} x_{i,j} E_{i,j}) \otimes (\sum_{k,l=1}^{f} y_{k,l} E_{k,l}) \\ &= \sum_{i,j=1}^{d} \sum_{k,l=1}^{f} x_{i,j} y_{k,l} (E_{i,j} \otimes E_{k,l}). \end{aligned} \tag{1.14}$$

But if $E_{i,j} \otimes E_{k,l}$ represents the (k,l)th position of the (i,j)th block of a matrix, then equation (1.14) says that the corresponding entry for $X \otimes Y$ should be $x_{i,j}y_{k,l}$, agreeing with the matrix definition.

We return from our brief detour to consider the center of $\operatorname{Com} X$. The *center* of an algebra A is

$$Z_A = \{a \in A \; : \; ab = ba \text{ for all } b \in A\}.$$

First we will compute the center of a matrix algebra. This result should be very reminiscent of Corollary 1.6.8 to Schur's lemma.

Proposition 1.7.6 *The center of* Mat_d *is*

$$Z_{\operatorname{Mat}_d} = \{cI_d \; : \; c \in \mathbb{C}\}.$$

Proof. Suppose that $C \in Z_{\operatorname{Mat}_d}$. Then, in particular,

$$CE_{i,i} = E_{i,i}C \tag{1.15}$$

for all i. But $CE_{i,i}$ (respectively, $E_{i,i}C$) is all zeros except for the ith column (respectively, row), which is the same as C's. Thus (1.15) implies that all off-diagonal elements of C must be 0. Similarly, if $i \neq j$, then

$$C(E_{i,j} + E_{j,i}) = (E_{i,j} + E_{j,i})C,$$

where the left (respectively, right) multiplication exchanges columns (respectively, rows) i and j of C. It follows that all the diagonal elements must be equal and so $C = cI_d$ for some $c \in \mathbb{C}$. Finally, all these matrices clearly commute with any other matrix, so we are done. ∎

Since we will be computing $Z_{\operatorname{Com} X}$ and the elements of the commutant algebra involve direct sums and tensor products, we will need to know how these operations behave under multiplication.

Lemma 1.7.7 *Suppose* $A, X \in \operatorname{Mat}_d$ *and* $B, Y \in \operatorname{Mat}_f$. *Then*

1. $(A \oplus B)(X \oplus Y) = AX \oplus BY$,

2. $(A \otimes B)(X \otimes Y) = AX \otimes BY$.

Proof. Both assertions are easy to prove, so we will do only the second. Suppose $A = (a_{i,j})$ and $X = (x_{i,j})$. Then

$$
\begin{aligned}
(A \otimes B)(X \otimes Y) &= (a_{i,j}B)(x_{i,j}Y) &&\text{(definition of } \otimes) \\
&= \left(\textstyle\sum_k a_{i,k}B \, x_{k,j}Y\right) &&\text{(block multiplication)} \\
&= \left(\left(\textstyle\sum_k a_{i,k}x_{k,j}\right)BY\right) &&\text{(distributivity)} \\
&= AX \otimes BY. &&\text{(definition of } \otimes) \; \blacksquare
\end{aligned}
$$

Now consider $C \in Z_{\operatorname{Com} X}$, where X and $\operatorname{Com} X$ are given by (1.12) and (1.13), respectively. So

$$CT = TC \text{ for all } T \in \operatorname{Com} X, \tag{1.16}$$

where $T = \oplus_{i=1}^{k}(M_{m_i} \otimes I_{d_i})$ and $C = \oplus_{i=1}^{k}(C_{m_i} \otimes I_{d_i})$. Computing the left-hand side, we obtain

$$
\begin{aligned}
CT &= (\oplus_{i=1}^{k}C_{m_i} \otimes I_{d_i})\,(\oplus_{i=1}^{k}M_{m_i} \otimes I_{d_i}) & \text{(definition of } C \text{ and } T) \\
&= \oplus_{i=1}^{k}(C_{m_i} \otimes I_{d_i})\,(M_{m_i} \otimes I_{d_i}) & \text{(Lemma 1.7.7, item 1)} \\
&= \oplus_{i=1}^{k}(C_{m_i}M_{m_i} \otimes I_{d_i}). & \text{(Lemma 1.7.7, item 2)}.
\end{aligned}
$$

Similarly,

$$TC = \oplus_{i=1}^{k}(M_{m_i}C_{m_i} \otimes I_{d_i}).$$

Thus equation (1.16) holds if and only if

$$C_{m_i}M_{m_i} = M_{m_i}C_{m_i} \text{ for all } M_{m_i} \in \text{Mat}_{m_i}.$$

But this just means that C_{m_i} is in the center of Mat_{m_i}, which, by Proposition 1.7.6, is equivalent to

$$C_{m_i} = c_i I_{m_i}$$

for some $c_i \in \mathbb{C}$. Hence

$$
\begin{aligned}
C &= \oplus_{i=1}^{k}c_i I_{m_i} \otimes I_{d_i} \\
&= \oplus_{i=1}^{k}c_i I_{m_i d_i} \\
&= \begin{pmatrix} c_1 I_{m_1 d_1} & 0 & \cdots & 0 \\ 0 & c_2 I_{m_2 d_2} & \cdots & 0 \\ \vdots & \vdots & \ddots & \vdots \\ 0 & 0 & \cdots & c_k I_{m_k d_k} \end{pmatrix},
\end{aligned}
$$

and all members of $Z_{\text{Com}\,X}$ have this form. Note that $\dim Z_{\text{Com}\,X} = k$.

For a concrete example, let

$$X = \begin{pmatrix} X^{(1)} & 0 & 0 \\ 0 & X^{(1)} & 0 \\ 0 & 0 & X^{(2)} \end{pmatrix} = 2X^{(1)} \oplus X^{(2)},$$

where $\deg X^{(1)} = 3$ and $\deg X^{(2)} = 4$. Then the matrices $T \in \text{Com}\,X$ look like

$$T = \left(\begin{array}{ccc|ccc|cccc} a & 0 & 0 & b & 0 & 0 & 0 & 0 & 0 & 0 \\ 0 & a & 0 & 0 & b & 0 & 0 & 0 & 0 & 0 \\ 0 & 0 & a & 0 & 0 & b & 0 & 0 & 0 & 0 \\ \hline c & 0 & 0 & d & 0 & 0 & 0 & 0 & 0 & 0 \\ 0 & c & 0 & 0 & d & 0 & 0 & 0 & 0 & 0 \\ 0 & 0 & c & 0 & 0 & d & 0 & 0 & 0 & 0 \\ \hline 0 & 0 & 0 & 0 & 0 & 0 & x & 0 & 0 & 0 \\ 0 & 0 & 0 & 0 & 0 & 0 & 0 & x & 0 & 0 \\ 0 & 0 & 0 & 0 & 0 & 0 & 0 & 0 & x & 0 \\ 0 & 0 & 0 & 0 & 0 & 0 & 0 & 0 & 0 & x \end{array} \right),$$

where $a, b, c, d, x \in \mathbb{C}$. The dimension is evidently

$$\dim(\text{Com}\,X) = m_1^2 + m_2^2 = 2^2 + 1^2 = 5.$$

The elements $C \in Z_{\mathrm{Com}\,X}$ are even simpler:

$$
C = \left(
\begin{array}{ccc|ccc|cccc}
a & 0 & 0 & 0 & 0 & 0 & 0 & 0 & 0 & 0 \\
0 & a & 0 & 0 & 0 & 0 & 0 & 0 & 0 & 0 \\
0 & 0 & a & 0 & 0 & 0 & 0 & 0 & 0 & 0 \\
\hline
0 & 0 & 0 & a & 0 & 0 & 0 & 0 & 0 & 0 \\
0 & 0 & 0 & 0 & a & 0 & 0 & 0 & 0 & 0 \\
0 & 0 & 0 & 0 & 0 & a & 0 & 0 & 0 & 0 \\
\hline
0 & 0 & 0 & 0 & 0 & 0 & x & 0 & 0 & 0 \\
0 & 0 & 0 & 0 & 0 & 0 & 0 & x & 0 & 0 \\
0 & 0 & 0 & 0 & 0 & 0 & 0 & 0 & x & 0 \\
0 & 0 & 0 & 0 & 0 & 0 & 0 & 0 & 0 & x
\end{array}
\right),
$$

where $a, x \in \mathbb{C}$. Here the dimension is the number of different irreducible components of X, in this case 2.

We summarize these results in the following theorem.

Theorem 1.7.8 *Let X be a matrix representation of G such that*

$$ X = m_1 X^{(1)} \oplus m_2 X^{(2)} \oplus \cdots \oplus m_k X^{(k)}, \qquad (1.17) $$

where the $X^{(i)}$ are inequivalent, irreducible and $\deg X^{(i)} = d_i$. Then

1. $\deg X = m_1 d_1 + m_2 d_2 + \cdots + m_k d_k$,

2. $\mathrm{Com}\,X = \{\oplus_{i=1}^{k}(M_{m_i} \otimes I_{d_i}) \; : \; M_{m_i} \in \mathrm{Mat}_{m_i} \text{ for all } i\}$,

3. $\dim(\mathrm{Com}\,X) = m_1^2 + m_2^2 + \cdots + m_k^2$,

4. $Z_{\mathrm{Com}\,X} = \{\oplus_{i=1}^{k} c_i I_{m_i d_i} \; : \; c_i \in \mathbb{C} \text{ for all } i\}$, *and*

5. $\dim Z_{\mathrm{Com}\,X} = k$. ∎

What happens if we try to apply Theorem 1.7.8 to a representation Y that is not decomposed into irreducibles? By the matrix version of Maschke's theorem (Corollary 1.5.4), Y is equivalent to a representation X of the form given in equation (1.17). But if $Y = RXR^{-1}$ for some fixed matrix R, then the map

$$ T \to RTR^{-1} $$

is an algebra isomorphism from $\mathrm{Com}\,X$ to $\mathrm{Com}\,Y$. Once the commutant algebras are isomorphic, it is easy to see that their centers are too. Hence Theorem 1.7.8 continues to hold with all set equalities replaced by isomorphisms.

There is also a module version of this result. We will use the multiplicity notation for G-modules in the same way it was used for matrices.

Theorem 1.7.9 *Let V be a G-module such that*

$$ V \cong m_1 V^{(1)} \oplus m_2 V^{(2)} \oplus \cdots \oplus m_k V^{(k)}, $$

where the $V^{(i)}$ are pairwise inequivalent irreducibles and $\dim V^{(i)} = d_i$. Then

1. $\dim V = m_1 d_1 + m_2 d_2 + \cdots + m_k d_k$,

2. $\operatorname{End} V \cong \oplus_{i=1}^{k} \operatorname{Mat}_{m_i}$,

3. $\dim(\operatorname{End} V) = m_1^2 + m_2^2 + \cdots + m_k^2$,

4. $Z_{\operatorname{End} V}$ *is isomorphic to the algebra of diagonal matrices of degree k, and*

5. $\dim Z_{\operatorname{End} V} = k$. ∎

The same methods can be applied to prove the following strengthening of Corollary 1.6.7 in the case where the field is $\mathbb{C}$.

Proposition 1.7.10 *Let V and W be G-modules with V irreducible. Then* $\dim \operatorname{Hom}(V, W)$ *is the multiplicity of V in W.* ∎

1.8   Group Characters

It turns out that much of the information contained in a representation can be distilled into one simple statistic: the traces of the corresponding matrices. This is the beautiful theory of group characters that will occupy us for the rest of this chapter.

Definition 1.8.1 *Let $X(g)$, $g \in G$, be a matrix representation. Then the* character of X *is*

$$\chi(g) = \operatorname{tr} X(g),$$

where tr *denotes the trace of a matrix. Otherwise put, χ is the map*

$$G \overset{\operatorname{tr} X}{\to} \mathbb{C}.$$

If V is a G-module, then its character *is the character of a matrix representation X corresponding to V.* ∎

Since there are many matrix representations corresponding to a single G-module, we should check that the module character is well-defined. But if X and Y both correspond to V, then $Y = TXT^{-1}$ for some fixed T. Thus, for all $g \in G$,

$$\operatorname{tr} Y(g) = \operatorname{tr} TX(g)T^{-1} = \operatorname{tr} X(g),$$

since trace is invariant under conjugation. Hence X and Y have the same character and our definition makes sense.

Much of the terminology we have developed for representations will be applied without change to the corresponding characters. Thus if X has character χ, we will say that χ is irreducible whenever X is, etc. Now let us turn to some examples.

Example 1.8.2 Suppose G is arbitrary and X is a degree 1 representation. Then the character $\chi(g)$ is just the sole entry of $X(g)$ for each $g \in G$. Such characters are called *linear characters*. ∎

Example 1.8.3 Suppose we consider the defining representation of S_n with its character χ^{def}. If we take $n = 3$, then we can compute the character values directly by taking the traces of the matrices in Example 1.2.4. The results are

$$\chi^{\text{def}}(\ (1)(2)(3)\ ) = 3, \qquad \chi^{\text{def}}(\ (1,2)(3)\ ) = 1, \qquad \chi^{\text{def}}(\ (1,3)(2)\ ) = 1,$$
$$\chi^{\text{def}}(\ (1)(2,3)\ ) = 1, \qquad \chi^{\text{def}}(\ (1,2,3)\ ) = 0, \qquad \chi^{\text{def}}(\ (1,3,2)\ ) = 0.$$

It is not hard to see that in general, if $\pi \in S_n$, then

$$\chi^{\text{def}}(\pi) \quad = \quad \text{the number of ones on the diagonal of } X(\pi)$$
$$= \quad \text{the number of fixedpoints of } \pi. \ ∎$$

Example 1.8.4 Let $G = \{g_1, g_2, \ldots, g_n\}$ and consider the regular representation with module $V = \mathbb{C}[\mathbf{G}]$ and character χ^{reg}. Now $X(\epsilon) = I_n$, so $\chi^{\text{reg}}(\epsilon) = |G|$.

To compute the character values for $g \neq \epsilon$, we will use the matrices arising from the standard basis $\mathcal{B} = \{\mathbf{g}_1, \mathbf{g}_2, \ldots, \mathbf{g}_n\}$. Now $X(g)$ is the permutation matrix for the action of g on $\mathcal{B}$, so $\chi^{\text{reg}}(g)$ is the number of fixedpoints for that action. But if $g\mathbf{g}_i = \mathbf{g}_i$ for any i, then we must have $g = \epsilon$, which is not the case; i.e., there are no fixedpoints if $g \neq \epsilon$. To summarize,

$$\chi^{\text{reg}}(g) = \begin{cases} |G| & \text{if } g = \epsilon, \\ 0 & \text{otherwise.} \end{cases} \ ∎$$

We now prove some elementary properties of characters.

Proposition 1.8.5 *Let X be a matrix representation of a group G of degree d with character χ.*

1. $\chi(\epsilon) = d$.

2. *If K is a conjugacy class of G, then*

$$g, h \in K \Rightarrow \chi(g) = \chi(h).$$

3. *If Y is a representation of G with character ψ, then*

$$X \cong Y \Rightarrow \chi(g) = \psi(g)$$

for all $g \in G$.

Proof. 1. Since $X(\epsilon) = I_d$,

$$\chi(\epsilon) = \operatorname{tr} I_d = d.$$

2. By hypothesis $g = khk^{-1}$, so

$$\chi(g) = \operatorname{tr} X(g) = \operatorname{tr} X(k)X(h)X(k)^{-1} = \operatorname{tr} X(h) = \chi(h).$$

3. This assertion just says that equivalent representations have the same character. We have already proved this in the remarks following the preceding definition of group characters. ∎

It is surprising that the converse of 3 is also true—i.e., if two representations have the same character, then they must be equivalent. This result (which is proved as Corollary 1.9.4, part 5) is the motivation for the paragraph with which we opened this section.

In the previous proposition, 2 says that characters are constant on conjugacy classes. Such functions have a special name.

Definition 1.8.6 A *class function* on a group G is a mapping $f : G \to \mathbb{C}$ such that $f(g) = f(h)$ whenever g and h are in the same conjugacy class. The set of all class functions on G is denoted by $R(G)$.

Clearly, the sums and scalar multiples of class functions are again class functions, so $R(G)$ is actually a vector space over $\mathbb{C}$. Also, $R(G)$ has a natural basis consisting of those functions that have the value 1 on a given conjugacy class and 0 elsewhere. Thus

$$\dim R(G) = \text{ number of conjugacy classes of } G. \qquad (1.18)$$

If K is a conjugacy class and χ is a character, we can define χ_K to be the value of the given character on the given class:

$$\chi_K = \chi(g)$$

for any $g \in K$. This brings us to the definition of the character table of a group.

Definition 1.8.7 Let G be a group. The *character table of G* is an array with rows indexed by the inequivalent irreducible characters of G and columns indexed by the conjugacy classes. The table entry in row χ and column K is χ_K:

$$
\begin{array}{c|ccc}
 & \cdots & K & \cdots \\
\hline
\vdots & & \vdots & \\
\chi & \cdots & \chi_K & \\
\vdots & &
\end{array}
$$

By convention, the first row corresponds to the trivial character, and the first column corresponds to the class of the identity, $K = \{\epsilon\}$. ∎

It is not clear that the character table is always finite: There might be an infinite number of irreducible characters of G. Fortunately, this turns out not to be the case. In fact, we will prove in Section 1.10 that the number of inequivalent irreducible representations of G is equal to the number of conjugacy classes, so the character table is always square. Let us examine some examples.

Example 1.8.8 If $G = C_n$, the cyclic group with n elements, then each element of C_n is in a conjugacy class by itself (as is true for any abelian group). Since there are n conjugacy classes, there must be n inequivalent irreducible representations of C_n. But we found n degree 1 representations in Example 1.2.3, and they are pairwise inequivalent, since they all have different characters (Proposition 1.8.5, part 3). So we have found all the irreducibles for C_n.

Since the representations are one-dimensional, they are equal to their corresponding characters. Thus the table we displayed on page 5 is indeed the complete character table for C_4. ∎

Example 1.8.9 Recall that a conjugacy class in $G = \mathcal{S}_n$ consists of all permutations of a given cycle type. In particular, for $\mathcal{S}_3$ we have three conjugacy classes,

$$K_1 = \{\epsilon\}, \qquad K_2 = \{(1,2),\ (1,3),\ (2,3)\}, \quad \text{and} \quad K_3 = \{(1,2,3),\ (1,3,2)\}.$$

Thus there are three irreducible representations of $\mathcal{S}_3$. We have met two of them, the trivial and sign representations. So this is as much as we know of the character table for $\mathcal{S}_3$:

	K_1	K_2	K_3
$\chi^{(1)}$	1	1	1
$\chi^{(2)}$	1	−1	1
$\chi^{(3)}$	?	?	?

We will be able to fill in the last line using character inner products. ∎

1.9 Inner Products of Characters

Next, we study the powerful tool of the character inner product. Taking inner products is a simple method for determining whether a representation is irreducible. This technique will also be used to prove that equality of characters implies equivalence of representations and to show that the number of irreducibles is equal to the number of conjugacy classes. First, however, we motivate the definition.

We can think of a character χ of a group $G = \{g_1, g_2, \ldots, g_n\}$ as a row vector of complex numbers:

$$\chi = (\chi(g_1), \chi(g_2), \ldots, \chi(g_n)).$$

If χ is irreducible, then this vector can be obtained from the character table by merely repeating the value for class K a total of $|K|$ times. For example, the first two characters for S_3 in the preceding table become

$$\chi^{(1)} = (1,1,1,1,1,1) \quad \text{and} \quad \chi^{(2)} = (1,-1,-1,-1,1,1).$$

We have the usual inner product on row vectors given by

$$(c_1, c_2, \ldots, c_n) \cdot (d_1, d_2, \ldots, d_n) = c_1 \overline{d_1} + c_2 \overline{d_2} + \cdots + c_n \overline{d_n},$$

where the bar stands for complex conjugation. Computing with our S_3 characters, it is easy to verify that

$$\chi^{(1)} \cdot \chi^{(1)} = \chi^{(2)} \cdot \chi^{(2)} = 6$$

and

$$\chi^{(1)} \cdot \chi^{(2)} = 0.$$

More trials with other irreducible characters—e.g., those of C_4—will lead the reader to conjecture that if $\chi^{(i)}$ and $\chi^{(j)}$ are irreducible characters of G, then

$$\chi^{(i)} \cdot \chi^{(j)} = \begin{cases} |G| & \text{if } i = j, \\ 0 & \text{if } i \neq j. \end{cases}$$

Dividing by $|G|$ for normality gives us one definition of the character inner product.

Definition 1.9.1 Let χ and ψ be any two functions from a group G to the complex numbers $\mathbb{C}$. The *inner product* of χ and ψ is

$$\langle \chi, \psi \rangle = \frac{1}{|G|} \sum_{g \in G} \chi(g) \overline{\psi(g)}. \quad \blacksquare$$

Now suppose V is a G-module with character ψ. We have seen, in the proof of Maschke's theorem, that there is an inner product on V itself that is invariant under the action of G. By picking an orthonormal basis for V, we obtain a matrix representation Y for ψ, where each $Y(g)$ is unitary; i.e.,

$$Y(g^{-1}) = Y(g)^{-1} = \overline{Y(g)}^t,$$

where t denotes transpose. So

$$\overline{\psi(g)} = \operatorname{tr} \overline{Y(g)} = \operatorname{tr} Y(g^{-1})^t = \operatorname{tr} Y(g^{-1}) = \psi(g^{-1}).$$

Substituting this into $\langle \cdot, \cdot \rangle$ yields another useful form of the inner product.

Proposition 1.9.2 Let χ and ψ be characters; then

$$\langle \chi, \psi \rangle = \frac{1}{|G|} \sum_{g \in G} \chi(g) \psi(g^{-1}). \quad \blacksquare \qquad (1.19)$$

When the field is arbitrary, equation (1.19) is taken as the *definition* of the inner product. In fact, for any two functions χ and ψ from G to a field, we can define

$$\langle \chi, \psi \rangle' = \frac{1}{|G|} \sum_{g \in G} \chi(g)\psi(g^{-1}),$$

but over the complex numbers this "inner product" is only a bilinear form. Of course, when restricted to characters we have $\langle \chi, \psi \rangle = \langle \chi, \psi \rangle'$. Also note that whenever χ and ψ are constant on conjugacy classes, we have

$$\langle \chi, \psi \rangle = \frac{1}{|G|} \sum_{K} |K| \chi_K \overline{\psi_K},$$

where the sum is over all conjugacy classes of G.

We can now prove that the irreducible characters are orthonormal with respect to the inner product $\langle \cdot, \cdot \rangle$.

Theorem 1.9.3 (Character Relations of the First Kind) *Let χ and ψ be irreducible characters of a group G. Then*

$$\langle \chi, \psi \rangle = \delta_{\chi, \psi}.$$

Proof. Suppose χ, ψ are the characters of matrix representations A, B of degrees d, f, respectively. We will be using Schur's lemma, and so a matrix must be found to fulfill the role of T in Corollary 1.6.6. Let $X = (x_{i,j})$ be a $d \times f$ matrix of indeterminates $x_{i,j}$ and consider the matrix

$$Y = \frac{1}{|G|} \sum_{g \in G} A(g) X B(g^{-1}) \tag{1.20}$$

We claim that $A(h)Y = YB(h)$ for all $h \in G$. Indeed,

$$
\begin{aligned}
A(h)YB(h)^{-1} &= \frac{1}{|G|} \sum_{g \in G} A(h)A(g)XB(g^{-1})B(h^{-1}) \\
&= \frac{1}{|G|} \sum_{g \in G} A(hg)XB(g^{-1}h^{-1}) \\
&= \frac{1}{|G|} \sum_{\substack{\tilde{g} \in G \\ \tilde{g} = hg}} A(\tilde{g})XB(\tilde{g}^{-1}) \\
&= Y,
\end{aligned}
$$

and our assertion is proved. Thus by Corollaries 1.6.6 and 1.6.8,

$$Y = \begin{cases} 0 & \text{if } A \not\cong B, \\ cI_d & \text{if } A \cong B. \end{cases} \tag{1.21}$$

Consider first the case where $\chi \neq \psi$, so that A and B must be inequivalent. Since this forces $y_{i,j} = 0$ for every element of Y, we can take the (i,j) entry of equation (1.20) to obtain

$$\frac{1}{|G|} \sum_{k,l} \sum_{g \in G} a_{i,k}(g) x_{k,l} b_{l,j}(g^{-1}) = 0$$

for all i, j. If this polynomial is to be zero, the coefficient of each $x_{k,l}$ must also be zero, so

$$\frac{1}{|G|} \sum_{g \in G} a_{i,k}(g) b_{l,j}(g^{-1}) = 0$$

for all i, j, k, l. Notice that this last equation can be more simply stated as

$$\langle a_{i,k}, b_{l,j} \rangle' = 0 \qquad \forall \, i, j, k, l, \tag{1.22}$$

since our definition of inner product applies to all functions from G to $\mathbb{C}$. Now,

$$\chi = \mathrm{tr}\, A = a_{1,1} + a_{2,2} + \cdots + a_{d,d}$$

and

$$\psi = \mathrm{tr}\, B = b_{1,1} + b_{2,2} + \cdots + b_{f,f},$$

so

$$\langle \chi, \psi \rangle = \langle \chi, \psi \rangle' = \sum_{i,j} \langle a_{i,i}, b_{j,j} \rangle' = 0$$

as desired.

Now suppose $\chi = \psi$. Since we are only interested in the character values, we might as well take $A = B$ also. By equation (1.21), there is a scalar $c \in \mathbb{C}$ such that $y_{i,j} = c\delta_{i,j}$. So, as in the previous paragraph, we have $\langle a_{i,k}, a_{l,j} \rangle' = 0$ as long as $i \neq j$. To take care of the case $i = j$, consider

$$\frac{1}{|G|} \sum_{g \in G} A(g) X A(g^{-1}) = c I_d$$

and take the trace on both sides:

$$\begin{aligned}
cd &= \mathrm{tr}\, c I_d \\
&= \frac{1}{|G|} \sum_{g \in G} \mathrm{tr}\, A(g) X A(g^{-1}) \\
&= \frac{1}{|G|} \sum_{g \in G} \mathrm{tr}\, X \\
&= \mathrm{tr}\, X.
\end{aligned}$$

Thus $y_{i,i} = c = \frac{1}{d} \mathrm{tr}\, X$, which can be rewritten as

$$\frac{1}{|G|} \sum_{k,l} \sum_{g \in G} a_{i,k}(g) x_{k,l} a_{l,i}(g^{-1}) = \frac{1}{d}(x_{1,1} + x_{2,2} + \cdots + x_{d,d}).$$

Equating coefficients of like monomials in this equation yields

$$\langle a_{i,k}, a_{l,i}\rangle' = \frac{1}{|G|} \sum_{g \in G} a_{i,k}(g) a_{l,i}(g^{-1}) = \frac{1}{d}\, \delta_{k,l}. \qquad (1.23)$$

It follows that

$$
\begin{aligned}
\langle \chi, \chi \rangle &= \sum_{i,j=1}^{d} \langle a_{i,i}, a_{j,j} \rangle' \\
&= \sum_{i=1}^{d} \langle a_{i,i}, a_{i,i} \rangle' \\
&= \sum_{i=1}^{d} \frac{1}{d} \\
&= 1,
\end{aligned}
$$

and the theorem is proved. ∎

Note that equations (1.22) and (1.23) give orthogonality relations for the matrix entries of the representations.

The character relations of the first kind have many interesting consequences.

Corollary 1.9.4 *Let X be a matrix representation of G with character χ. Suppose*

$$X \cong m_1 X^{(1)} \oplus m_2 X^{(2)} \oplus \cdots \oplus m_k X^{(k)},$$

where the $X^{(i)}$ are pairwise inequivalent irreducibles with characters $\chi^{(i)}$.

1. $\chi = m_1 \chi^{(1)} + m_2 \chi^{(2)} + \cdots + m_k \chi^{(k)}$.

2. $\langle \chi, \chi^{(j)} \rangle = m_j$ *for all j.*

3. $\langle \chi, \chi \rangle = m_1^2 + m_2^2 + \cdots + m_k^2$.

4. X *is irreducible if and only if $\langle \chi, \chi \rangle = 1$.*

5. *Let Y be another matrix representation of G with character ψ. Then*

$$X \cong Y \text{ if and only if } \chi(g) = \psi(g)$$

   for all $g \in G$.

Proof. 1. Using the fact that the trace of a direct sum is the sum of the traces, we see that

$$\chi = \operatorname{tr} X = \operatorname{tr} \bigoplus_{i=1}^{k} m_i X^{(i)} = \sum_{i=1}^{k} m_i \chi^{(i)}.$$

2. We have, by the previous theorem,

$$\langle \chi, \chi^{(j)} \rangle = \langle \sum_i m_i \chi^{(i)}, \chi^{(j)} \rangle = \sum_i m_i \langle \chi^{(i)}, \chi^{(j)} \rangle = m_j.$$

3. By another application of Theorem 1.9.3:

$$\langle \chi, \chi \rangle = \langle \sum_i m_i \chi^{(i)}, \sum_j m_j \chi^{(j)} \rangle = \sum_{i,j} m_i m_j \langle \chi^{(i)}, \chi^{(j)} \rangle = \sum_i m_i^2$$

4. The assertion that X is irreducible implies that $\langle \chi, \chi \rangle = 1$ is just part of the orthogonality relations already proved. For the converse, suppose that

$$\langle \chi, \chi \rangle = \sum_i m_i^2 = 1.$$

Then there must be exactly one index j such that $m_j = 1$ and all the rest of the m_i must be zero. But then $X = X^{(j)}$, which is irreducible by assumption.

5. The forward implication was proved as part 3 of Proposition 1.8.5. For the other direction, let $Y \cong \oplus_{i=1}^k n_i X^{(i)}$. There is no harm in assuming that the X and Y expansions both contain the same irreducibles: Any irreducible found in one but not the other can be inserted with multiplicity 0. Now $\chi = \psi$, so $\langle \chi, \chi^{(i)} \rangle = \langle \psi, \chi^{(i)} \rangle$ for all i. But then, by part 2 of this corollary, $m_i = n_i$ for all i. Thus the two direct sums are equivalent—i.e., $X \cong Y$. ∎

As an example of how these results are applied in practice, we return to the defining representation of $\mathcal{S}_n$. To simplify matters, note that both $\pi, \pi^{-1} \in \mathcal{S}_n$ have the same cycle type and are thus in the same conjugacy class. So if χ is a character of $\mathcal{S}_n$, then $\chi(\pi) = \chi(\pi^{-1})$, since characters are constant on conjugacy classes. It follows that the inner product formula for $\mathcal{S}_n$ can be rewritten as

$$\langle \chi, \psi \rangle = \frac{1}{n!} \sum_{\pi \in \mathcal{S}_n} \chi(\pi) \psi(\pi). \tag{1.24}$$

Example 1.9.5 Let $G = \mathcal{S}_3$ and consider $\chi = \chi^{\mathrm{def}}$. Let $\chi^{(1)}, \chi^{(2)}, \chi^{(3)}$ be the three irreducible characters of $\mathcal{S}_3$, where the first two are the trivial and sign characters, respectively. By Maschke's theorem, we know that

$$\chi = m_1 \chi^{(1)} + m_2 \chi^{(2)} + m_3 \chi^{(3)}.$$

Furthermore, we can use equation (1.24) and part 2 of Corollary 1.9.4 to compute m_1 and m_2 (character values for $\chi = \chi^{\mathrm{def}}$ were found in Example 1.8.3):

$$m_1 = \langle \chi, \chi^{(1)} \rangle = \frac{1}{3!} \sum_{\pi \in \mathcal{S}_3} \chi(\pi) \chi^{(1)}(\pi) = \frac{1}{6}(3\cdot1+1\cdot1+1\cdot1+1\cdot1+0\cdot1+0\cdot1) = 1,$$

$$m_2 = \langle \chi, \chi^{(2)} \rangle = \frac{1}{3!} \sum_{\pi \in \mathcal{S}_3} \chi(\pi) \chi^{(2)}(\pi) = \frac{1}{6}(3\cdot1-1\cdot1-1\cdot1-1\cdot1+0\cdot1+0\cdot1) = 0.$$

Thus
$$\chi = \chi^{(1)} + m_3\chi^{(3)}.$$

In fact, we already knew that the defining character contained a copy of the trivial one. This was noted when we decomposed the corresponding matrices as $X = A \oplus B$, where A was the matrix of the trivial representation (see page 13). The exciting news is that the B matrices correspond to one or more copies of the mystery character $\chi^{(3)}$. These matrices turned out to be

$$B(\epsilon) = \begin{pmatrix} 1 & 0 \\ 0 & 1 \end{pmatrix}, \qquad B((1,2)) = \begin{pmatrix} -1 & -1 \\ 0 & 1 \end{pmatrix},$$

$$B((1,3)) = \begin{pmatrix} 1 & 0 \\ -1 & -1 \end{pmatrix}, \qquad B((2,3)) = \begin{pmatrix} 0 & 1 \\ 1 & 0 \end{pmatrix},$$

$$B((1,2,3)) = \begin{pmatrix} -1 & -1 \\ 1 & 0 \end{pmatrix}, \qquad B((1,3,2)) = \begin{pmatrix} 0 & 1 \\ -1 & -1 \end{pmatrix}.$$

If we let ψ be the corresponding character, then

$$\psi(\epsilon) = 2,$$
$$\psi((1,2)) = \psi((1,3)) = \psi((2,3)) = 0,$$
$$\psi((1,2,3)) = \psi((1,3,2)) = -1.$$

If ψ is irreducible, then $m_3 = 1$ and we have found $\chi^{(3)}$. (If not, then ψ, being of degree 2, must contain two copies of $\chi^{(3)}$.) But part 4 of Corollary 1.9.4 makes it easy to determine irreducibility; merely compute:

$$\langle \psi, \psi \rangle = \frac{1}{6}(2^2 + 0^2 + 0^2 + 0^2 + (-1)^2 + (-1)^2) = 1.$$

We have found the missing irreducible. The complete character table for S_3 is thus

	K_1	K_2	K_3
$\chi^{(1)}$	1	1	1
$\chi^{(2)}$	1	-1	1
$\chi^{(3)}$	2	0	-1

In general, the defining module for S_n, $V = \mathbb{C}\{1, 2, \ldots, n\}$, always has $W = \mathbb{C}\{1 + 2 + \cdots + n\}$ as a submodule. If $\chi^{(1)}$ and $\chi^{\perp}$ are the characters corresponding to W and $W^{\perp}$, respectively, then $V = W \oplus W^{\perp}$. This translates to

$$\chi^{\text{def}} = \chi^{(1)} + \chi^{\perp}$$

on the character level. We already know that χ^{def} counts fixedpoints and that $\chi^{(1)}$ is the trivial character. Thus

$$\chi^{\perp}(\pi) = \text{ (number of fixedpoints of } \pi) - 1$$

is also a character of S_n. In fact, $\chi^{\perp}$ is irreducible, although that is not obvious from the previous discussion. ■

1.10   Decomposition of the Group Algebra

We now apply the machinery we have developed to the problem of decomposing the group algebra into irreducibles. In the process, we determine the number of inequivalent irreducible representations of any group.

Let G be a group with group algebra $\mathbb{C}[\mathbf{G}]$ and character $\chi = \chi^{\text{reg}}$. By Maschke's theorem (Theorem 1.5.3), we can write

$$\mathbb{C}[\mathbf{G}] = \bigoplus_i m_i V^{(i)}, \tag{1.25}$$

where the $V^{(i)}$ run over all pairwise inequivalent irreducibles (and only a finite number of the m_i are nonzero).

What are the multiplicities m_i? If $V^{(i)}$ has character $\chi^{(i)}$, then, by part 2 of Corollary 1.9.4,

$$m_i = \langle \chi, \chi^{(i)} \rangle = \frac{1}{|G|} \sum_{g \in G} \chi(g) \chi^{(i)}(g^{-1}).$$

But we computed the character of the regular representation in Example 1.8.4, and it vanished for $g \neq \epsilon$ with $\chi(\epsilon) = |G|$. Plugging in the preceding values, we obtain

$$m_i = \frac{1}{|G|} \chi(\epsilon) \chi^{(i)}(\epsilon) = \dim V^{(i)} \tag{1.26}$$

by Proposition 1.8.5, part 1. Hence every irreducible G-module occurs in $\mathbb{C}[\mathbf{G}]$ with multiplicity equal to its dimension. In particular, they all appear at least once, so the list of inequivalent irreducibles must be finite (since the group algebra has finite dimension). We record these results, among others about the decomposition of $\mathbb{C}[\mathbf{G}]$, as follows.

Proposition 1.10.1 *Let G be a finite group and suppose $\mathbb{C}[\mathbf{G}] = \oplus_i m_i V^{(i)}$, where the $V^{(i)}$ form a complete list of pairwise inequivalent irreducible G-modules. Then*

1. $m_i = \dim V^{(i)}$,

2. $\sum_i (\dim V^{(i)})^2 = |G|$, *and*

3. *The number of $V^{(i)}$ equals the number of conjugacy classes of G.*

Proof. Part 1 is proved above, and from it, part 2 follows by taking dimensions in equation (1.25).

For part 3, recall from Theorem 1.7.9 that

$$\text{number of } V^{(i)} = \dim Z_{\text{End } \mathbb{C}[\mathbf{G}]}.$$

What do the elements of End $\mathbb{C}[\mathbf{G}]$ look like? Given any $\mathbf{v} \in \mathbb{C}[\mathbf{G}]$, define the map $\phi_{\mathbf{v}} : \mathbb{C}[\mathbf{G}] \to \mathbb{C}[\mathbf{G}]$ to be right multiplication by $\mathbf{v}$, i.e.,

$$\phi_{\mathbf{v}}(\mathbf{w}) = \mathbf{w}\mathbf{v}.$$

for all $\mathbf{w} \in \mathbb{C}[\mathbf{G}]$. It is easy to verify that $\phi_\mathbf{v} \in \operatorname{End} \mathbb{C}[\mathbf{G}]$. In fact, these are the only elements of $\operatorname{End} \mathbb{C}[\mathbf{G}]$ because we claim that $\mathbb{C}[\mathbf{G}] \cong \operatorname{End} \mathbb{C}[\mathbf{G}]$ as vector spaces. To see this, consider $\phi : \mathbb{C}[\mathbf{G}] \rightarrow \operatorname{End} \mathbb{C}[\mathbf{G}]$ given by

$$\mathbf{v} \overset{\phi}{\rightarrow} \phi_\mathbf{v}.$$

Proving that ϕ is linear is not hard. To show injectivity, we compute its kernel. If $\phi_\mathbf{v}$ is the zero map, then

$$0 = \phi_\mathbf{v}(\epsilon) = \epsilon \mathbf{v} = \mathbf{v}.$$

For surjectivity, suppose $\theta \in \operatorname{End} \mathbb{C}[\mathbf{G}]$ and consider $\theta(\epsilon)$, which is some vector $\mathbf{v}$. It follows that $\theta = \phi_\mathbf{v}$, because given any $g \in G$,

$$\theta(\mathbf{g}) = \theta(g\epsilon) = g\theta(\epsilon) = g\mathbf{v} = \mathbf{g}\mathbf{v} = \phi_\mathbf{v}(\mathbf{g}),$$

and two linear transformations that agree on a basis agree everywhere. On the algebra level, our map ϕ is an anti-isomorphism, since it reverses the order of multiplication: $\phi_\mathbf{v}\phi_\mathbf{w} = \phi_{\mathbf{w}\mathbf{v}}$ for all $\mathbf{v}, \mathbf{w} \in \mathbb{C}[\mathbf{G}]$. Thus ϕ induces an anti-isomorphism of the centers of $\mathbb{C}[\mathbf{G}]$ and $\operatorname{End} \mathbb{C}[\mathbf{G}]$, so that

$$\text{number of } V^{(i)} = \dim Z_{\mathbb{C}[\mathbf{G}]}.$$

To find out what the center of the group algebra looks like, consider any $\mathbf{z} = c_1\mathbf{g}_1 + \cdots + c_n\mathbf{g}_n \in Z_{\mathbb{C}[\mathbf{G}]}$, where the g_i are in G. Now for all $h \in G$, we have $\mathbf{z}\mathbf{h} = \mathbf{h}\mathbf{z}$, or $\mathbf{z} = \mathbf{h}\mathbf{z}\mathbf{h}^{-1}$, which can be written out as

$$c_1\mathbf{g}_1 + \cdots + c_n\mathbf{g}_n = c_1\mathbf{h}\mathbf{g}_1\mathbf{h}^{-1} + \cdots + c_n\mathbf{h}\mathbf{g}_n\mathbf{h}^{-1}.$$

But as h takes on all possible values in G, hg_1h^{-1} runs over the conjugacy class of g_1. Since $\mathbf{z}$ remains invariant, all members of this class must have the same scalar coefficient c_1. Thus if G has k conjugacy classes $K_1, \ldots, K_k$ and we let

$$\mathbf{z}_i = \sum_{g \in K_i} \mathbf{g}$$

for $i = 1, \ldots, k$, then we have shown that any $\mathbf{z} \in Z_{\mathbb{C}[\mathbf{G}]}$ can be written as

$$\mathbf{z} = \sum_{i=1}^{k} d_i\mathbf{z}_i.$$

Similar considerations show that the converse holds: Any linear combination of the $\mathbf{z}_i$ is in the center of $\mathbb{C}[\mathbf{G}]$. Finally, we note that the set $\{\mathbf{z}_1, \ldots, \mathbf{z}_k\}$ forms a basis for $Z_{\mathbb{C}[\mathbf{G}]}$. We have already shown that they span. They must also be linearly independent, since they are sums over pairwise disjoint subsets of the basis $\{\mathbf{g} : g \in G\}$ of $\mathbb{C}[\mathbf{G}]$. Hence

$$\text{number of conjugacy classes} = \dim Z_{\mathbb{C}[\mathbf{G}]} = \text{number of } V^{(i)}$$

as desired. ∎

As a first application of this proposition, we derive a slightly deeper relationship between the characters and class functions.

Proposition 1.10.2 *The irreducible characters of a group G form an orthonormal basis for the space of class functions $R(G)$.*

Proof. Since the irreducible characters are orthonormal with respect to the bilinear form $\langle \cdot, \cdot \rangle$ on $R(G)$ (Theorem 1.9.3), they are linearly independent. But part 3 of Proposition 1.10.1 and equation (1.18) show that we have $\dim R(G)$ such characters. Thus they are a basis. ∎

Knowing that the character table is square permits us to derive orthogonality relations for its columns as a companion to those for the rows.

Theorem 1.10.3 (Character Relations of the Second Kind) *Let K, L be conjugacy classes of G. Then*

$$\sum_\chi \chi_K \overline{\chi_L} = \frac{|G|}{|K|} \delta_{K,L},$$

where the sum is over all irreducible characters of G.

Proof. If χ and ψ are irreducible characters, then the character relations of the first kind yield

$$\langle \chi, \psi \rangle = \frac{1}{|G|} \sum_K |K| \, \chi_K \, \overline{\psi_K} = \delta_{\chi, \psi},$$

where the sum is over all conjugacy classes of G. But this says that the modified character table

$$U = \left(\sqrt{|K|/|G|} \, \chi_K \right)$$

has orthonormal rows. Hence U, being square, is a unitary matrix and has orthonormal columns. The theorem follows. ∎

As a third application of these ideas, we can now give an alternative method for finding the third line of the character table for $\mathcal{S}_3$ that does not involve actually producing the corresponding representation. Let the three irreducible characters be $\chi^{(1)}$, $\chi^{(2)}$, and $\chi^{(3)}$, where the first two are the trivial and sign characters, respectively. If d denotes the dimension of the corresponding module for $\chi^{(3)}$, then by Proposition 1.10.1, part 2,

$$1^2 + 1^2 + d^2 = |\mathcal{S}_3| = 6.$$

Thus $\chi^{(3)}(\epsilon) = d = 2$. To find the value of $\chi^{(3)}$ on any other permutation, we use the orthogonality relations of the second kind. For example, to compute $x = \chi^{(3)}(\,(1,2)\,)$,

$$0 = \sum_{i=1}^3 \chi^{(i)}(\epsilon)\overline{\chi^{(i)}(\,(1,2)\,)} = 1 \cdot 1 + 1(-1) + 2\overline{x},$$

so $x = 0$.

1.11 Tensor Products Again

Suppose we have representations of groups G and H and wish to construct a representation of the product group $G \times H$. It turns out that we can use the tensor product introduced in Section 1.7 for this purpose. In fact, all the irreducible representations of $G \times H$ can be realized as tensor products of irreducibles for the individual groups. Proving this provides another application of the theory of characters.

Definition 1.11.1 Let G and H have matrix representations X and Y, respectively. The *tensor product representation*, $X \otimes Y$, assigns to each $(g, h) \in G \times H$ the matrix

$$(X \otimes Y)(g, h) = X(g) \otimes Y(h). \blacksquare$$

We must verify that this is indeed a representation. While we are at it, we might as well compute its character.

Theorem 1.11.2 *Let X and Y be matrix representations for G and H, respectively.*

1. *Then $X \otimes Y$ is a representation of $G \times H$.*

2. *If X, Y and $X \otimes Y$ have characters denoted by χ, ψ, and $\chi \otimes \psi$, respectively, then*

$$(\chi \otimes \psi)(g, h) = \chi(g)\psi(h)$$

for all $(g, h) \in G \times H$.

Proof. 1. We verify the two conditions defining a representation. First of all,

$$(X \otimes Y)(\epsilon, \epsilon) = X(\epsilon) \otimes Y(\epsilon) = I \otimes I = I.$$

Secondly, if $(g, h), (g', h') \in G \times H$, then using Lemma 1.7.7, part 2,

$$
\begin{aligned}
(X \otimes Y)((g, h) \cdot (g', h')) &= (X \otimes Y)(gg', hh') \\
&= X(gg') \otimes Y(hh') \\
&= X(g)X(g') \otimes Y(h)Y(h') \\
&= (X(g) \otimes Y(h)) \cdot (X(g') \otimes Y(h')) \\
&= (X \otimes Y)(g, h) \cdot (X \otimes Y)(g', h').
\end{aligned}
$$

2. Note that for any matrices A and B,

$$\operatorname{tr} A \otimes B = \operatorname{tr}(a_{i,j}B) = \sum_i a_{i,i} \operatorname{tr} B = \operatorname{tr} A \operatorname{tr} B.$$

Thus

$$(\chi \otimes \psi)(g, h) = \operatorname{tr}(X(g) \otimes Y(h)) = \operatorname{tr} X(g) \operatorname{tr} Y(h) = \chi(g)\psi(h). \blacksquare$$

There is a module-theoretic way to view the tensor product of representations. Let V be a G-module and W be an H-module. Then we can turn the vector space $V \otimes W$ into a $G \times H$-module by defining

$$(g, h)(\mathbf{v} \otimes \mathbf{w}) = (g\mathbf{v}) \otimes (h\mathbf{w})$$

and linearly extending the action as $\mathbf{v}$ and $\mathbf{w}$ run through bases of V and W, respectively. It is not hard to show that this definition satisfies the module axioms and is independent of the choice of bases. Furthermore, if V and W correspond to matrix representations X and Y via the basis vectors consisting of $\mathbf{v}$'s and $\mathbf{w}$'s, then $V \otimes W$ is a module for $X \otimes Y$ in the basis composed of $\mathbf{v} \otimes \mathbf{w}$'s.

Now we show how the irreducible representations of G and H completely determine those of $G \times H$.

Theorem 1.11.3 *Let G and H be groups.*

1. *If X and Y are irreducible representations of G and H, respectively, then $X \otimes Y$ is an irreducible representation of $G \times H$.*

2. *If $X^{(i)}$ and $Y^{(j)}$ are complete lists of inequivalent irreducible representations for G and H, respectively, then $X^{(i)} \otimes Y^{(j)}$ is a complete list of inequivalent irreducible $G \times H$-modules.*

Proof. 1. If ϕ is any character, then we know (Corollary 1.9.4, part 4) that the corresponding representation is irreducible if and only if $\langle \phi, \phi \rangle = 1$. Letting X and Y have characters χ and ψ, respectively, we have

$$
\begin{aligned}
\langle \chi \otimes \psi, \chi \otimes \psi \rangle &= \frac{1}{|G \times H|} \sum_{(g,h) \in G \times H} (\chi \otimes \psi)(g, h)(\chi \otimes \psi)(g^{-1}, h^{-1}) \\
&= \left[\frac{1}{|G|} \sum_{g \in G} \chi(g)\chi(g^{-1}) \right] \left[\frac{1}{|H|} \sum_{h \in H} \psi(h)\psi(h^{-1}) \right] \\
&= \langle \chi, \chi \rangle \langle \psi, \psi \rangle \\
&= 1 \cdot 1 \\
&= 1.
\end{aligned}
$$

2. Let $X^{(i)}$ and $Y^{(j)}$ have characters $\chi^{(i)}$ and $\psi^{(j)}$, respectively. Then as in the proof of 1, we can show that

$$\langle \chi^{(i)} \otimes \psi^{(j)}, \chi^{(k)} \otimes \psi^{(l)} \rangle = \langle \chi^{(i)}, \chi^{(k)} \rangle \langle \psi^{(j)}, \psi^{(l)} \rangle = \delta_{i,k}\delta_{j,l}.$$

Thus from Corollary 1.9.4, part 3, we see that the $\chi^{(i)} \otimes \psi^{(j)}$ are pairwise inequivalent.

To prove that the list is complete, it suffices to show that the number of such representations is the number of conjugacy classes of $G \times H$ (Proposition 1.10.1, part 3). But this is the number of conjugacy classes of G times the number of conjugacy classes of H, which is the number of $X^{(i)} \otimes Y^{(j)}$. ∎

1.12 Restricted and Induced Representations

Given a group G with a subgroup H, is there a way to get representations of G from those of H or vice versa? We can answer these questions in the affirmative using the operations of restriction and induction. In fact, we have already seen an example of the latter, namely, the coset representation of Example 1.3.5.

Definition 1.12.1 Consider $H \leq G$ and a matrix representation X of G. The *restriction of X to H, $X{\downarrow}_H^G$*, is given by

$$X{\downarrow}_H^G (h) = X(h)$$

for all $h \in H$. If X has character χ, then denote the character of $X{\downarrow}_H^G$ by $\chi{\downarrow}_H^G$. ∎

It is trivial to verify that $X{\downarrow}_H^G$ is actually a representation of H. If the group G is clear from context, we will drop it in the notation and merely write $X{\downarrow}_H$. Note that even though X may be an irreducible representation of G, $X{\downarrow}_H$ need not be irreducible on H.

The process of moving from a representation of H to one of G is a bit more involved. This construction, called induction, is due to Frobenius. Suppose Y is a matrix representation of H. We might try to obtain a representation of G by merely defining Y to be zero outside of H. This will not work because singular matrices are not allowed. However, there is a way to remedy the situation.

Definition 1.12.2 Consider $H \leq G$ and fix a transversal $t_1, \ldots, t_l$ for the left cosets of H, i.e., $G = t_1 H \uplus \cdots \uplus t_l H$, where $\uplus$ denotes disjoint union. If Y is a representation of H, then the corresponding *induced representation* $Y{\uparrow}_H^G$ assigns to each $g \in G$ the block matrix

$$Y{\uparrow}_H^G (g) = (Y(t_i^{-1} g t_j)) = \begin{pmatrix} Y(t_1^{-1} g t_1) & Y(t_1^{-1} g t_2) & \cdots & Y(t_1^{-1} g t_l) \\ Y(t_2^{-1} g t_1) & Y(t_2^{-1} g t_2) & \cdots & Y(t_2^{-1} g t_l) \\ \vdots & \vdots & \ddots & \vdots \\ Y(t_l^{-1} g t_1) & Y(t_l^{-1} g t_2) & \cdots & Y(t_l^{-1} g t_l) \end{pmatrix},$$

where $Y(g)$ is the zero matrix if $g \notin H$. ∎

We will abbreviate $Y{\uparrow}_H^G$ to $Y{\uparrow}^G$ if no confusion will result; this notation applies to the corresponding characters as well. It is not obvious that $Y{\uparrow}_H^G$ is actually a representation of G, but we will postpone that verification until after we have looked at an example.

As usual, let $G = S_3$ and consider $H = \{\epsilon, (2,3)\}$ with the transversal $G = H \uplus (1,2)H \uplus (1,3)H$ as in Example 1.6.3. Let $Y = 1$ be the trivial representation of H and consider $X = 1{\uparrow}^G$. Calculating the first row of the matrix for the transposition (1,2) yields

$$
\begin{aligned}
Y(\,\epsilon^{-1}(1,2)\epsilon\,) &= Y(\,(1,2)\,) = 0, && \text{(since } (1,2)\notin H) \\
Y(\,\epsilon^{-1}(1,2)(1,2)\,) &= Y(\,\epsilon\,) = 1, && \text{(since } \epsilon \in H) \\
Y(\,\epsilon^{-1}(1,2)(1,3)\,) &= Y(\,(1,3,2)\,) = 0. && \text{(since } (1,3,2)\notin H)
\end{aligned}
$$

Continuing in this way, we obtain

$$
X(\,(1,2)\,) = \begin{pmatrix} 0 & 1 & 0 \\ 1 & 0 & 0 \\ 0 & 0 & 1 \end{pmatrix}.
$$

The matrices for the coset representation are beginning to appear again. This is not an accident, as the next proposition shows.

Proposition 1.12.3 *Let $H \le G$ have transversal $\{t_1,\ldots,t_l\}$ with cosets $\mathcal{H} = \{t_1H,\ldots,t_lH\}$. Then the matrices of $1\uparrow_H^G$ are identical with those of G acting on the basis $\mathcal{H}$ for the coset module $\mathbb{C}\mathcal{H}$.*

Proof. Let the matrices for $1\uparrow^G$ and $\mathbb{C}\mathcal{H}$ be $X = (x_{i,j})$ and $Z = (z_{i,j})$, respectively. Both arrays contain only zeros and ones. Finally, for any $g \in G$,

$$
\begin{aligned}
x_{i,j}(g) = 1 \;&\Leftrightarrow\; t_i^{-1}gt_j \in H \\
&\Leftrightarrow\; gt_jH = t_iH \\
&\Leftrightarrow\; z_{i,j}(g) = 1. \;\blacksquare
\end{aligned}
$$

Thus $\mathbb{C}\mathcal{H}$ is a module for $1\uparrow_H^G$.

It is high time that we verified that induced representations are well-defined.

Theorem 1.12.4 *Suppose $H \le G$ has transversal $\{t_1,\ldots,t_l\}$ and let Y be a matrix representation of H. Then $X = Y\uparrow_H^G$ is a representation of G.*

Proof. Analogous to the case where Y is the trivial representation, we prove that $X(g)$ is always a block permutation matrix; i.e., every row and column contains exactly one nonzero block $Y(t_i^{-1}gt_j)$. Consider the first column (the other cases being similar). It suffices to show that there is a unique element of H on the list $t_1^{-1}gt_1, t_2^{-1}gt_1,\ldots,t_l^{-1}gt_1$. But $gt_1 \in t_iH$ for exactly one of the t_i in our transversal, and so $t_i^{-1}gt_1 \in H$ is the element we seek.

We must first verify that $X(\epsilon)$ is the identity matrix, but that follows directly from the definition of induction.

Also, we need to show that $X(g)X(h) = X(gh)$ for all $g,h \in G$. Considering the (i,j) block on both sides, it suffices to prove

$$
\sum_k Y(t_i^{-1}gt_k)Y(t_k^{-1}ht_j) = Y(t_i^{-1}ght_j).
$$

For ease of notation, let $a_k = t_i^{-1}gt_k$, $b_k = t_k^{-1}ht_j$, and $c = t_i^{-1}ght_j$. Note that $a_kb_k = c$ for all k and that the sum can be rewritten

$$
\sum_k Y(a_k)Y(b_k) \overset{?}{=} Y(c).
$$

Now the proof breaks into two cases.

If $Y(c) = 0$, then $c \notin H$, and so either $a_k \notin H$ or $b_k \notin H$ for all k. Thus $Y(a_k)$ or $Y(b_k)$ is zero for each k, which forces the sum to be zero as well.

If $Y(c) \neq 0$, then $c \in H$. Let m be the unique index such that $a_m \in H$. Thus $b_m = a_m^{-1}c \in H$, and so

$$\sum_k Y(a_k)Y(b_k) = Y(a_m)Y(b_m) = Y(a_m b_m) = Y(c),$$

completing the proof. ∎

It should be noted that induction, like restriction, does not preserve irreducibility. It might also seem that an induced representation depends on the transversal and not just on the subgroup chosen. However, this is an illusion.

Proposition 1.12.5 *Consider $H \leq G$ and a matrix representation Y of H. Let $\{t_1, \ldots, t_l\}$ and $\{s_1, \ldots, s_l\}$ be two transversals for H giving rise to representation matrices X and Z, respectively, for $Y{\uparrow}^G$. Then X and Z are equivalent.*

Proof. Let χ, ψ, and ϕ be the characters of X, Y, and Z, respectively. Then it suffices to show that $\chi = \phi$ (Corollary 1.9.4, part 5). Now

$$\chi(g) = \sum_i \operatorname{tr} Y(t_i^{-1}gt_i) = \sum_i \psi(t_i^{-1}gt_i), \qquad (1.27)$$

where $\psi(g) = 0$ if $g \notin H$. Similarly,

$$\phi(g) = \sum_i \psi(s_i^{-1}gs_i).$$

Since the t_i and s_i are both transversals, we can permute subscripts if necessary to obtain $t_i H = s_i H$ for all i. Now $t_i = s_i h_i$, where $h_i \in H$ for all i, and so

$$t_i^{-1}gt_i = h_i^{-1}s_i^{-1}gs_ih_i.$$

Thus $t_i^{-1}gt_i \in H$ if and only if $s_i^{-1}gs_i \in H$, and when both lie in H, they are in the same conjugacy class. It follows that $\psi(t_i^{-1}gt_i) = \psi(s_i^{-1}gs_i)$, since ψ is constant on conjugacy classes of H and zero outside. Hence the sums for χ and ϕ are the same. ∎

We next derive a useful formula for the character of an induced representation. Let H, ψ, and the t_i be as in the preceding proposition. Then $\psi(t_i^{-1}gt_i) = \psi(h^{-1}t_i^{-1}gt_ih)$ for any $h \in H$, so equation (1.27) can be rewritten as

$$\psi{\uparrow}^G(g) = \frac{1}{|H|} \sum_i \sum_{h \in H} \psi(h^{-1}t_i^{-1}gt_ih).$$

But as h runs over H and the t_i run over the transversal, the product t_ih runs over all the elements of G exactly once. Thus we arrive at the identity

$$\psi{\uparrow}^G(g) = \frac{1}{|H|} \sum_{x \in G} \psi(x^{-1}gx). \qquad (1.28)$$

This formula will permit us to prove the celebrated reciprocity law of Frobenius, which relates inner products of restricted and induced characters.

Theorem 1.12.6 (Frobenius Reciprocity) *Let $H \leq G$ and suppose that ψ and χ are characters of H and G, respectively. Then*

$$\langle \psi \uparrow^G, \chi \rangle = \langle \psi, \chi \downarrow_H \rangle,$$

where the left inner product is calculated in G and the right one in H.

Proof. We have the following string of equalities:

$$
\begin{aligned}
\langle \psi \uparrow^G, \chi \rangle &= \frac{1}{|G|} \sum_{g \in G} \psi \uparrow^G (g) \chi(g^{-1}) \\
&= \frac{1}{|G||H|} \sum_{x \in G} \sum_{g \in G} \psi(x^{-1}gx) \chi(g^{-1}) \quad \text{(equation (1.28))} \\
&= \frac{1}{|G||H|} \sum_{x \in G} \sum_{y \in G} \psi(y) \chi(xy^{-1}x^{-1}) \quad \text{(let } y = x^{-1}gx\text{)} \\
&= \frac{1}{|G||H|} \sum_{x \in G} \sum_{y \in G} \psi(y) \chi(y^{-1}) \quad \text{(χ constant on G's classes)} \\
&= \frac{1}{|H|} \sum_{y \in G} \psi(y) \chi(y^{-1}) \quad \text{(x constant in the sum)} \\
&= \frac{1}{|H|} \sum_{y \in H} \psi(y) \chi(y^{-1}) \quad \text{(ψ is zero outside H)} \\
&= \langle \psi, \chi \downarrow_H \rangle. \quad \blacksquare
\end{aligned}
$$

1.13 Exercises

1. An *inversion* in $\pi = x_1, x_2, \ldots, x_n \in S_n$ (one-line notation) is a pair x_i, x_j such that $i < j$ and $x_i > x_j$. Let inv π be the number of inversions of π.

   (a) Show that if π can be written as a product of k transpositions, then $k \equiv \text{inv } \pi \pmod 2$.

   (b) Use part (a) to show that the sign of π is well-defined.

2. If group G acts on a set S and $s \in S$, then the *stabilizer of s* is $G_s = \{g \in G : gs = s\}$. The *orbit of s* is $\mathcal{O}_s = \{gs : g \in G\}$.

   (a) Prove that G_s is a subgroup of G.

   (b) Find a bijection between cosets of G/G_s and elements of $\mathcal{O}_s$.

   (c) Show that $|\mathcal{O}_s| = |G|/|G_s|$ and use this to derive formula (1.1) for $|K_g|$.

3. Let G act on S with corresponding permutation representation $\mathbb{C}S$. Prove the following.

   (a) The matrices for the action of G in the standard basis are permutation matrices.

   (b) If the character of this representation is χ and $g \in G$, then

$$\chi(g) = \text{ the number of fixed points of } g \text{ acting on } S.$$

4. Let G be an abelian group. Find all inequivalent irreducible representations of G. *Hint:* Use the fundamental theorem of abelian groups.

5. If X is a matrix representation of a group G, then its *kernel* is the set $N = \{g \in G \ : \ X(g) = I\}$. A representation is *faithful* if it is one-to-one.

   (a) Show that N is a normal subgroup of G and find a condition on N equivalent to the representation being faithful.

   (b) Suppose X has character χ and degree d. Prove that $g \in N$ if and only if $\chi(g) = d$. *Hint:* Show that $\chi(g)$ is a sum of roots of unity.

   (c) Show that for the coset representation, $N = \cap_i g_i H g_i^{-1}$, where the g_i are the transversal.

   (d) For each of the following representations, under what conditions are they faithful: trivial, regular, coset, sign for S_n, defining for S_n, degree 1 for C_n?

   (e) Define a function Y on the group G/N by $Y(gN) = X(g)$ for $gN \in G/N$.

       i. Prove that Y is a well-defined faithful representation of G/N.

       ii. Show that Y is irreducible if and only if X is.

       iii. If X is the coset representation for a normal subgroup H of G, what is the corresponding representation Y?

6. It is possible to reverse the process of part (e) in the previous exercise. Let N be any normal subgroup of G and let Y be a representation of G/N. Define a function on G by $X(g) = Y(gN)$.

   (a) Prove that X is a representation of G. We say that X has been *lifted* from the representation Y of G/N.

   (b) Show that if Y is faithful, then X has kernel N.

   (c) Show that X is irreducible if and only if Y is.

7. Let X be a reducible matrix representation with block form given by equation (1.4). Let V be a module for X with submodule W corresponding to A. Consider the quotient vector space $V/W = \{\mathbf{v}+W \ : \ \mathbf{v} \in V\}$. Show that V/W is a G-module with corresponding matrices $C(g)$. Furthermore, show that we have

$$V \cong W \oplus (V/W).$$

8. Let V be a vector space. Show that the following properties will hold in V if and only if a related property holds for a basis of V.

    (a) V is a G-module.

    (b) The map $\theta : V \to W$ is a G-homomorphism.

    (c) The inner product $\langle \cdot, \cdot \rangle$ on V is G-invariant.

9. Why won't replacing the inner products in the proof of Maschke's theorem by bilinear forms give a demonstration valid over any field? Give a correct proof over an arbitrary field as follows. Assume that X is a reducible matrix representation of the form (1.4).

    (a) Write out the equations obtained by equating blocks in $X(gh) = X(g)X(h)$. What interpretation can you give to the equations obtained from the upper left and lower right blocks?

    (b) Use part (a) to show that

$$TX(g)T^{-1} = \begin{pmatrix} A(g) & 0 \\ 0 & C(g) \end{pmatrix},$$

    where $T = \begin{pmatrix} I & D \\ 0 & I \end{pmatrix}$ and $D = \frac{1}{|G|} \sum_{g \in G} A(g^{-1})B(g)$.

10. Verify that the map $X : \mathbf{R}^+ \to GL_2$ given in the example at the end of Section 1.5 is a representation and that the subspace W is invariant.

11. Find $H \le S_n$ and a set of tabloids S such that $\mathbb{C}\mathcal{H} \cong \mathbb{C}S \cong \mathbb{C}\{1, 2, \ldots, n\}$.

12. Let X be an irreducible matrix representation of G. Show that if $g \in Z_G$ (the center of G), then $X(g) = cI$ for some scalar c.

13. Let $\{X_1, X_2, \ldots, X_n\} \subseteq GL_d$ be a subgroup of commuting matrices. Show that these matrices are simultaneously diagonalizable using representation theory.

14. Prove the following converse of Schur's lemma. Let X be a representation of G over $\mathbb{C}$ with the property that only scalar multiples cI commute with $X(g)$ for all $g \in G$. Prove that X is irreducible.

15. Let X and Y be representations of G. The *inner tensor product*, $X \hat{\otimes} Y$, assigns to each $g \in G$ the matrix

$$(X \hat{\otimes} Y)(g) = X(g) \otimes Y(g).$$

    (a) Verify that $X \hat{\otimes} Y$ is a representation of G.

    (b) Show that if X, Y, and $X \hat{\otimes} Y$ have characters denoted by χ, ψ, and $\chi \hat{\otimes} \psi$, respectively, then $(\chi \hat{\otimes} \psi)(g) = \chi(g)\psi(g)$.

    (c) Find a group with irreducible representations X and Y such that $X \hat{\otimes} Y$ is *not* irreducible.

(d) However, prove that if X is of degree 1 and Y is irreducible, then so is $X \hat{\otimes} Y$.

16. Construct the character table of $\mathcal{S}_4$. You may find the lifting process of Exercise 6 and the inner tensor products of Exercise 15 helpful.

17. Let D_n be the group of symmetries (rotations and reflections) of a regular n-gon. This group is called a *dihedral group*.

(a) Show that the abstract group with generators ρ, τ subject to the relations
$$\rho^n = \tau^2 = \epsilon \quad \text{and} \quad \rho\tau = \tau\rho^{-1}$$
is isomorphic to D_n.

(b) Conclude that every element of D_n is uniquely expressible as $\tau^i \rho^j$, where $0 \le i \le 1$ and $0 \le j \le n - 1$.

(c) Find the conjugacy classes of D_n.

(d) Find all the inequivalent irreducible *representations* of D_n. *Hint:* Use the fact that C_n is a normal subgroup of D_n.

18. Show that induction is transitive as follows. Suppose we have groups $G \ge H \ge K$ and a matrix representation X of K. Then
$$X{\uparrow}_K^G \cong (X{\uparrow}_K^H){\uparrow}_H^G .$$

Chapter 2

Representations of the Symmetric Group

In this chapter we construct all the irreducible representations of the symmetric group. We know that the number of such representations is equal to the number of conjugacy classes (Proposition 1.10.1), which in the case of $\mathcal{S}_n$ is the number of partitions of n. It may not be obvious how to associate an irreducible with each partition $\lambda = (\lambda_1, \lambda_2, \ldots, \lambda_l)$, but it is easy to find a corresponding subgroup $\mathcal{S}_\lambda$ that is an isomorphic copy of $\mathcal{S}_{\lambda_1} \times \mathcal{S}_{\lambda_2} \times \cdots \times \mathcal{S}_{\lambda_l}$ inside $\mathcal{S}_n$. We can now produce the right number of representations by inducing the trivial representation on each $\mathcal{S}_\lambda$ up to $\mathcal{S}_n$.

If M^λ is a module for $1 \uparrow_{\mathcal{S}_\lambda}^{\mathcal{S}_n}$, then it is too much to expect that these modules will all be irreducible. However, we will be able to find an ordering $\lambda^{(1)}, \lambda^{(2)}, \ldots$ of all partitions of n with the following nice property. The first module $M^{\lambda^{(1)}}$ will be irreducible; call it $S^{\lambda^{(1)}}$. Next, $M^{\lambda^{(2)}}$ will contain only copies of $S^{\lambda^{(1)}}$ plus a single copy of a new irreducible $S^{\lambda^{(2)}}$. In general, $M^{\lambda^{(i)}}$ will decompose into some $S^{\lambda^{(k)}}$ for $k < i$ and a unique new irreducible $S^{\lambda^{(i)}}$ called the ith Specht module. Thus the matrix giving the multiplicities for expressing $M^{\lambda^{(i)}}$ as a direct sum of the $S^{\lambda^{(j)}}$ will be lower triangular with ones down the diagonal. This immediately makes it easy to compute the irreducible characters of $\mathcal{S}_n$. We can also explicitly describe the Specht modules themselves.

Much of the material in this chapter can be found in the monograph of James [Jam 78], where the reader will find a more extensive treatment.

2.1 Young Subgroups, Tableaux, and Tabloids

Our objective in this section is to build the modules M^λ. First, however, we will introduce some notation and definitions for partitions.

If $\lambda = (\lambda_1, \lambda_2, \ldots, \lambda_l)$ is a partition of n, then we write $\lambda \vdash n$. We also use the notation $|\lambda| = \sum_i \lambda_i$, so that a partition of n satisfies $|\lambda| = n$. We can visualize λ as follows.

Definition 2.1.1 Suppose $\lambda = (\lambda_1, \lambda_2, \ldots, \lambda_l) \vdash n$. The *Ferrers diagram*, or *shape*, of λ is an array of n dots having l left-justified rows with row i containing λ_i dots for $1 \le i \le l$. ∎

The dot in row i and column j has coordinates (i, j), as in a matrix. Boxes (also called *cells*) are often used in place of dots. As an example, the partition $\lambda = (3, 3, 2, 1)$ has Ferrers diagram

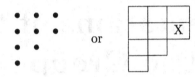

where the box in the $(2, 3)$ position has an X in it. The reader should be aware that certain authors, notably Francophones, write their Ferrers diagrams as if they were in a Cartesian coordinate system, with the first row along the x-axis, the next along the line $y = 1$, etc. With this convention, our example partition would be drawn as

but we will stick to "English" notation in this book. Also, we will use the symbol λ to stand for both the partition and its shape.

Now we wish to associate with λ a subgroup of $\mathcal{S}_n$. Recall that if T is any set, then $\mathcal{S}_T$ is the set of permutations of T.

Definition 2.1.2 Let $\lambda = (\lambda_1, \lambda_2, \ldots, \lambda_l) \vdash n$. Then the corresponding *Young subgroup* of $\mathcal{S}_n$ is

$$\mathcal{S}_\lambda = \mathcal{S}_{\{1,2,\ldots,\lambda_1\}} \times \mathcal{S}_{\{\lambda_1+1, \lambda_1+2, \ldots, \lambda_1+\lambda_2\}} \times \cdots \times \mathcal{S}_{\{n-\lambda_l+1, n-\lambda_l+2, \ldots, n\}}.$$ ∎

These subgroups are named in honor of the Reverend Alfred Young, who was among the first to construct the irreducible representations of $\mathcal{S}_n$ [You 27, You 29]. For example,

$$\begin{aligned} \mathcal{S}_{(3,3,2,1)} &= \mathcal{S}_{\{1,2,3\}} \times \mathcal{S}_{\{4,5,6\}} \times \mathcal{S}_{\{7,8\}} \times \mathcal{S}_{\{9\}} \\ &\cong \mathcal{S}_3 \times \mathcal{S}_3 \times \mathcal{S}_2 \times \mathcal{S}_1, \end{aligned}$$

In general, $\mathcal{S}_{(\lambda_1, \lambda_2, \ldots, \lambda_l)}$ and $\mathcal{S}_{\lambda_1} \times \mathcal{S}_{\lambda_2} \times \cdots \times \mathcal{S}_{\lambda_l}$ are isomorphic as groups.

Now consider the representation $1 \uparrow_{\mathcal{S}_\lambda}^{\mathcal{S}_n}$. If $\pi_1, \pi_2, \ldots, \pi_k$ is a transversal for $\mathcal{S}_\lambda$, then by Proposition 1.12.3 the vector space

$$V^\lambda = \mathbb{C}\{\pi_1 \mathcal{S}_\lambda,\ \pi_2 \mathcal{S}_\lambda,\ \ldots,\ \pi_k \mathcal{S}_\lambda\}$$

is a module for our induced representation. The cosets can be thought of geometrically as certain arrays in the following way.

Definition 2.1.3 Suppose $\lambda \vdash n$. A *Young tableau of shape* λ, is an array t obtained by replacing the dots of the Ferrers diagram of λ with the numbers $1, 2, \ldots, n$ bijectively. ∎

Let $t_{i,j}$ stand for the entry of t in position (i, j). A Young tableau of shape λ is also called a λ-*tableau* and denoted by t^λ. Alternatively, we write $\operatorname{sh} t = \lambda$. Clearly, there are $n!$ Young tableaux for any shape $\lambda \vdash n$. To illustrate, if

$$\lambda = (2, 1) = \begin{matrix} \bullet & \bullet \\ \bullet & \end{matrix},$$

then a list of all possible tableaux of shape λ is

$$t : \quad \begin{matrix} 1 & 2 \\ 3 & \end{matrix}, \quad \begin{matrix} 2 & 1 \\ 3 & \end{matrix}, \quad \begin{matrix} 1 & 3 \\ 2 & \end{matrix}, \quad \begin{matrix} 3 & 1 \\ 2 & \end{matrix}, \quad \begin{matrix} 2 & 3 \\ 1 & \end{matrix}, \quad \begin{matrix} 3 & 2 \\ 1 & \end{matrix}.$$

The first tableau above has entries $t_{1,1} = 1$, $t_{1,2} = 2$, and $t_{2,1} = 3$.

Young tabloids were mentioned in Example 1.6.3. We can now introduce them formally as equivalence classes of tableaux.

Definition 2.1.4 Two λ-tableaux t_1 and t_2 are *row equivalent*, $t_1 \sim t_2$, if corresponding rows of the two tableaux contain the same elements. A *tabloid of shape* λ, or λ-*tabloid*, is then

$$\{t\} = \{t_1 | t_1 \sim t\}$$

where $\operatorname{sh} t = \lambda$. ∎

As before, we will use lines between the rows of an array to indicate that it is a tabloid. So if

$$t = \begin{matrix} 1 & 2 \\ 3 & \end{matrix},$$

then

$$\{t\} = \left\{ \begin{matrix} 1 & 2 \\ 3 & \end{matrix}, \begin{matrix} 2 & 1 \\ 3 & \end{matrix} \right\} = \begin{matrix} \overline{1 \quad 2} \\ 3 \end{matrix}.$$

If $\lambda = (\lambda_1, \lambda_2, \ldots, \lambda_l) \vdash n$, then the number of tableaux in any given equivalence class is $\lambda_1! \lambda_2! \cdots \lambda_l! \overset{\text{def}}{=} \lambda!$. Thus the number of λ-tabloids is just $n!/\lambda!$.

Now $\pi \in \mathcal{S}_n$ acts on a tableau $t = (t_{i,j})$ of shape $\lambda \vdash n$ as follows:

$$\pi t = (\pi(t_{i,j}))$$

For example,

$$(1, 2, 3) \begin{matrix} 1 & 2 \\ 3 & \end{matrix} = \begin{matrix} 2 & 3 \\ 1 & \end{matrix}.$$

This induces an action on tabloids by letting

$$\pi\{t\} = \{\pi t\}.$$

The reader should check that this is well defined—i.e., independent of the choice of t. Finally, this action gives rise, in the usual way, to an $\mathcal{S}_n$-module.

Definition 2.1.5 Suppose $\lambda \vdash n$. Let

$$M^\lambda = \mathbb{C}\{\{t_1\}, \ldots, \{t_k\}\},$$

where $\{t_1\}, \ldots, \{t_k\}$ is a complete list of λ-tabloids. Then M^λ is called the *permutation module corresponding to* λ. ∎

Since we are considering only row equivalence classes, we could list the rows of M^λ in any order and produce an isomorphic module. Thus M^μ is defined for any composition (ordered partition) μ. As will be seen in the next examples, we have already met three of these modules.

Example 2.1.6 If $\lambda = (n)$, then

$$M^{(n)} = \mathbb{C}\{\ \overline{1\ 2\ \cdots\ n}\ \}$$

with the trivial action. ∎

Example 2.1.7 Now consider $\lambda = (1^n)$. Each equivalence class $\{t\}$ consists of a single tableau, and this tableau can be identified with a permutation in one-line notation (by taking the transpose, if you wish). Since the action of $\mathcal{S}_n$ is preserved,

$$M^{(1^n)} \cong \mathbb{C}\mathcal{S}_n,$$

and the regular representation presents itself. ∎

Example 2.1.8 Finally, if $\lambda = (n-1, 1)$, then each λ-tabloid is uniquely determined by the element in its second row, which is a number from 1 to n. As in Example 1.6.3, this sets up a module isomorphism

$$M^{(n-1,1)} \cong \mathbb{C}\{1, 2, \ldots, n\},$$

so we have recovered the defining representation. ∎

Example 2.1.9 Let us compute the full set of characters of these modules when $n = 3$. In this case the only partitions are $\lambda = (3)$, $(2, 1)$, and (1^3). From the previous examples, the modules M^λ correspond to the trivial, defining, and regular representations of $\mathcal{S}_3$, respectively. These character values have all been computed in Chapter 1. Denote the character of M^λ by ϕ^λ and represent the conjugacy class of $\mathcal{S}_3$ corresponding to μ by K_μ. Then we have the following table of characters:

	$K_{(1^3)}$	$K_{(2,1)}$	$K_{(3)}$
$\phi^{(3)}$	1	1	1
$\phi^{(2,1)}$	3	1	0
$\phi^{(1^3)}$	6	0	0 ∎

The M^λ enjoy the following general property of modules.

Definition 2.1.10 Any G-module M is *cyclic* if there is a $\mathbf{v} \in M$ such that

$$M = \mathbb{C}G\mathbf{v},$$

where $G\mathbf{v} = \{g\mathbf{v}|g \in G\}$. In this case we say that M is *generated by* $\mathbf{v}$. ∎

Since any λ-tabloid can be taken to any other tabloid of the same shape by some permutation, M^λ is cyclic. We summarize in the following proposition.

Proposition 2.1.11 *If* $\lambda \vdash n$, *then* M^λ *is cyclic, generated by any given* λ-*tabloid. In addition,* $\dim M^\lambda = n!/\lambda!$, *the number of* λ-*tabloids.* ∎

What is the connection between our permutation modules and those obtained by inducing up from a Young subgroup? One can think of

$$S_\lambda = S_{\{1,2,\ldots,\lambda_1\}} \times S_{\{\lambda_1+1,\lambda_1+2,\ldots,\lambda_1+\lambda_2\}} \times \cdots \times S_{\{n-\lambda_l+1,n-\lambda_l+2,\ldots,n\}}$$

as being modeled by the tabloid

$$\{t^\lambda\} = \begin{array}{|ccccc|}
\hline
1 & & 2 & \cdots & \lambda_1 \\
\hline
\lambda_1+1 & & \lambda_1+2 & \cdots & \lambda_1+\lambda_2 \\
\hline
& & \vdots & & \\
\hline
n-\lambda_l+1 & \cdots & & n & \\
\hline
\end{array}.$$

The fact that S_λ contains all permutations of a given interval of integers is mirrored by the fact that the order of these integers is immaterial in t^λ (since they all occur in the same row). Thus the coset πS_λ corresponds in some way to the tabloid $\{\pi t^\lambda\}$. To be more precise, consider Theorem 2.1.12.

Theorem 2.1.12 *Consider* $\lambda \vdash n$ *with Young subgroup* S_λ *and tabloid* $\{t^\lambda\}$, *as before. Then* $V^\lambda = \mathbb{C}S_n S_\lambda$ *and* $M^\lambda = \mathbb{C}S_n\{t^\lambda\}$ *are isomorphic as* S_n-*modules.*

Proof. Let $\pi_1, \pi_2, \ldots, \pi_k$ be a transversal for S_λ. Define a map

$$\theta : V^\lambda \to M^\lambda$$

by $\theta(\pi_i S_\lambda) = \{\pi_i t^\lambda\}$ for $i = 1, \ldots, k$ and linear extension. It is not hard to verify that θ is the desired S_n-isomorphism of modules. ∎

2.2 Dominance and Lexicographic Ordering

We need to find an ordering of partitions λ such that the M^λ have the nice property of the introduction to this chapter. In fact, we consider two important orderings on partitions of n, one of which is only a partial order.

Definition 2.2.1 If A is a set, then a *partial order on A* is a relation $\leq$ such that

1. $a \leq a$,

2. $a \leq b$ and $b \leq a$ implies $a = b$, and

3. $a \leq b$ and $b \leq c$ implies $a \leq c$

for all $a, b, c \in A$. In this case we say that $(A, \leq)$ is a *partially ordered set*, or *poset* for short. We also write $b \geq a$ for $a \leq b$ as well as $a < b$ (or $b > a$) for $a \leq b$ and $a \neq b$.

If, in addition, for every pair of elements $a, b \in A$ we have either $a \leq b$ or $b \leq a$, then $\leq$ is a *total order* and $(A, \leq)$ is a *totally ordered set*. ∎

The three laws for a partial order are called reflexivity, antisymmetry, and transitivity, respectively. Pairs of elements $a, b \in A$ such that neither $a \leq b$ nor $b \leq a$ holds are said to be *incomparable*. One of the simplest examples of a poset is the *Boolean algebra* $B_n = (A, \subseteq)$, where A is all subsets of $\{1, 2, \ldots, n\}$ ordered by inclusion. In B_n, the subsets $\{1\}$ and $\{2\}$ are incomparable. The set $\{0, 1, \ldots, n\}$ with the normal ordering is an example of a totally ordered set called the *n-chain* and denoted by C_n.

The particular partial order in which we are interested is the following.

Definition 2.2.2 Suppose $\lambda = (\lambda_1, \lambda_2, \ldots, \lambda_l)$ and $\mu = (\mu_1, \mu_2, \ldots, \mu_m)$ are partitions of n. Then λ *dominates* μ, written $\lambda \trianglerighteq \mu$, if

$$\lambda_1 + \lambda_2 + \cdots + \lambda_i \geq \mu_1 + \mu_2 + \cdots + \mu_i$$

for all $i \geq 1$. If $i > l$ (respectively, $i > m$), then we take λ_i (respectively, μ_i) to be zero. ∎

Intuitively, λ is greater than μ in the dominance order if the Ferrers diagram of λ is short and fat but the one for μ is long and skinny. For example, when $n = 6$, then $(3, 3) \trianglerighteq (2, 2, 1, 1)$ since $3 \geq 2$, $3 + 3 \geq 2 + 2$, etc. However, $(3, 3)$ and $(4, 1, 1)$ are incomparable since $3 \leq 4$ but $3 + 3 \geq 4 + 1$.

Any partially ordered set can be visualized using a Hasse diagram.

Definition 2.2.3 If $(A, \leq)$ is a poset and $b, c \in A$, then we say that b *is covered by c* (or c *covers* b), written $b \prec c$ (or $c \succ b$), if $b < c$ and there is no $d \in A$ with $b < d < c$. The *Hasse diagram of A* consists of vertices representing the elements of A with an arrow from vertex b up to vertex c if b is covered by c. ∎

The Hasse diagram for the partitions of 6 ordered by dominance is given next.

$$(6)$$
$$\uparrow$$
$$(5,1)$$
$$\uparrow$$
$$(4,2)$$

$$(3^2) \qquad\qquad (4,1^2)$$

$$(3,2,1)$$

$$(3,1^3) \qquad\qquad (2^3)$$

$$(2^2,1^2)$$
$$\uparrow$$
$$(2,1^4)$$
$$\uparrow$$
$$(1^6)$$

The fundamental lemma concerning the dominance order is as follows.

Lemma 2.2.4 (Dominance Lemma for Partitions) *Let t^λ and s^μ be tableaux of shape λ and μ, respectively. If, for each index i, the elements of row i of s^μ are all in different columns in t^λ, then $\lambda \trianglerighteq \mu$.*

Proof. By hypothesis, we can sort the entries in each column of t^λ so that the elements of rows $1, 2, \ldots, i$ of s^μ all occur in the first i rows of t^λ. Thus

$$
\begin{aligned}
\lambda_1 + \lambda_2 + \cdots + \lambda_i \;=\;& \text{number of elements in the first } i \text{ rows of } t^\lambda \\
\geq\;& \text{number of elements of } s^\mu \text{ in the first } i \text{ rows of } t^\lambda \\
=\;& \mu_1 + \mu_2 + \cdots + \mu_i. \quad\blacksquare
\end{aligned}
$$

The second ordering on partitions is the one that would be given to them in a dictionary.

Definition 2.2.5 Let $\lambda = (\lambda_1, \lambda_2, \ldots, \lambda_l)$ and $\mu = (\mu_1, \mu_2, \ldots, \mu_m)$ be partitions of n. Then $\lambda < \mu$ in *lexicographic order* if, for some index i,

$$\lambda_j = \mu_j \text{ for } j < i \text{ and } \lambda_i < \mu_i. \quad\blacksquare$$

This is a total ordering on partitions. For partitions of 6 we have

$$(1^6) < (2,1^4) < (2^2,1^2) < (2^3) < (3,1^3) < (3,2,1)$$
$$< (3^2) < (4,1^2) < (4,2) < (5,1) < (6).$$

The lexicographic order is a *refinement* of the dominance order in the sense of the following proposition.

Proposition 2.2.6 *If $\lambda, \mu \vdash n$ with $\lambda \trianglerighteq \mu$, then $\lambda \geq \mu$.*

Proof. If $\lambda \neq \mu$, then find the first index i where they differ. Thus $\sum_{j=1}^{i-1} \lambda_j = \sum_{j=1}^{i-1} \mu_j$ and $\sum_{j=1}^{i} \lambda_j > \sum_{j=1}^{i} \mu_j$ (since $\lambda \triangleright \mu$). So $\lambda_i > \mu_i$. ∎

It turns out that we will want to list the M^λ in *dual* lexicographic order—i.e., starting with the largest partition and working down. (By convention, the conjugacy classes are listed in the character table in the usual dictionary order so as to start with (1^n), which is the class of the identity.) Note that our first module will then be $M^{(n)}$, which is one-dimensional with the trivial action. Thus we have an irreducible module to start with, as promised.

In most of our theorems, however, we will use the dominance ordering in order to obtain the stronger conclusion that $\lambda \trianglerighteq \mu$ rather than just $\lambda > \mu$.

2.3 Specht Modules

We now construct all the irreducible modules of $\mathcal{S}_n$. These are the so-called Specht modules, S^λ.

Any tableau naturally determines certain isomorphic copies of Young subgroups in $\mathcal{S}_n$.

Definition 2.3.1 Suppose that the tableau t has rows $R_1, R_2, \ldots, R_l$ and columns $C_1, C_2, \ldots, C_k$. Then

$$R_t = \mathcal{S}_{R_1} \times \mathcal{S}_{R_2} \times \cdots \times \mathcal{S}_{R_l}$$

and

$$C_t = \mathcal{S}_{C_1} \times \mathcal{S}_{C_2} \times \cdots \times \mathcal{S}_{C_k}$$

are the *row-stabilizer* and *column-stabilizer* of t, respectively. ∎

If we take

$$t = \begin{matrix} 4 & 1 & 2 \\ 3 & 5 & \end{matrix} , \qquad (2.1)$$

then

$$R_t = \mathcal{S}_{\{1,2,4\}} \times \mathcal{S}_{\{3,5\}}$$

and

$$C_t = \mathcal{S}_{\{3,4\}} \times \mathcal{S}_{\{1,5\}} \times \mathcal{S}_{\{2\}}.$$

Note that our equivalence classes can be expressed as $\{t\} = R_t t$. In addition, these groups are associated with certain elements of $\mathbb{C}[\mathcal{S}_n]$. In general, given a subset $H \subseteq \mathcal{S}_n$, we can form the group algebra sums

$$H^+ = \sum_{\pi \in H} \pi$$

and

$$H^- = \sum_{\pi \in H} \text{sgn}(\pi)\pi.$$

(Here, the group algebra will be acting on the permutation modules M^λ. Thus the elements of $\mathbb{C}[S_n]$ are not playing the roles of vectors and are not set in boldface.) For a tableau t, the element R_t^+ is already implicit in the corresponding tabloid by the remark at the end of the previous paragraph. However, we will also need to make use of

$$\kappa_t \stackrel{\text{def}}{=} C_t^- = \sum_{\pi \in C_t} \text{sgn}(\pi)\pi.$$

Note that if t has columns $C_1, C_2, \ldots, C_k$, then κ_t factors as

$$\kappa_t = \kappa_{C_1} \kappa_{C_2} \cdots \kappa_{C_k}.$$

Finally, we can pass from t to an element of the module M^λ by the following definition.

Definition 2.3.2 If t is a tableau, then the associated *polytabloid* is

$$e_t = \kappa_t\{t\}. \quad \blacksquare$$

To illustrate these concepts using the tableau t in (2.1), we compute

$$\kappa_t = (\epsilon - (3,4))(\epsilon - (1,5)),$$

and so

$$e_t = \begin{array}{|c c c|} \hline 4 & 1 & 2 \\ \hline 3 & 5 \\ \hline \end{array} - \begin{array}{|c c c|} \hline 3 & 1 & 2 \\ \hline 4 & 5 \\ \hline \end{array} - \begin{array}{|c c c|} \hline 4 & 5 & 2 \\ \hline 3 & 1 \\ \hline \end{array} + \begin{array}{|c c c|} \hline 3 & 5 & 2 \\ \hline 4 & 1 \\ \hline \end{array}.$$

The next lemma describes what happens to the various objects defined previously in passing from t to πt.

Lemma 2.3.3 *Let t be a tableau and π be a permutation. Then*

1. $R_{\pi t} = \pi R_t \pi^{-1}$,

2. $C_{\pi t} = \pi C_t \pi^{-1}$,

3. $\kappa_{\pi t} = \pi \kappa_t \pi^{-1}$,

4. $e_{\pi t} = \pi e_t$.

Proof. 1. We have the following list of equivalent statements:

$$
\begin{aligned}
\sigma \in R_{\pi t} &\longleftrightarrow \sigma\{\pi t\} = \{\pi t\} \\
&\longleftrightarrow \pi^{-1}\sigma\pi\{t\} = \{t\} \\
&\longleftrightarrow \pi^{-1}\sigma\pi \in R_t \\
&\longleftrightarrow \sigma \in \pi R_t \pi^{-1}.
\end{aligned}
$$

The proofs of parts 2 and 3 are similar to that of part 1.

4. We have

$$e_{\pi t} = \kappa_{\pi t}\{\pi t\} = \pi\kappa_t\pi^{-1}\{\pi t\} = \pi\kappa_t\{t\} = \pi e_t. \ \blacksquare$$

One can think of this lemma, for example part 1, as follows. If t has entries $t_{i,j}$, then πt has entries $\pi t_{i,j}$. Thus an element of the row-stabilizer of πt may be constructed by first applying π^{-1} to obtain the tableau t, then permuting the elements within each row of t, and finally reapplying π to restore the correct labels. Alternatively, we have just shown that if $\sigma \in R_{\pi t}$, then $\pi^{-1}\sigma\pi \in R_t$ and the following diagram commutes:

$$
\begin{array}{ccc}
 & \sigma & \\
\pi t & \longrightarrow & \pi t_1 \\
\pi^{-1} \downarrow & & \uparrow \ \pi \\
t & \longrightarrow & t_1 \\
 & \pi^{-1}\sigma\pi &
\end{array}
$$

Finally, we are in a position to define the Specht modules.

Definition 2.3.4 For any partition λ, the corresponding *Specht module*, S^λ, is the submodule of M^λ spanned by the polytabloids e_t, where t is of shape λ. $\blacksquare$

Because of part 4 of Lemma 2.3.3, we have the following.

Proposition 2.3.5 *The S^λ are cyclic modules generated by any given polytabloid.* $\blacksquare$

Let us look at some examples.

Example 2.3.6 Suppose $\lambda = (n)$. Then $e_{1\ 2\ \cdots\ n} = \boxed{1\ 2\ \cdots\ n}$ is the only polytabloid and $S^{(n)}$ carries the trivial representation. This is, of course, the only possibility, since $S^{(n)}$ is a submodule of $M^{(n)}$ where $\mathcal{S}_n$ acts trivially (Example 2.1.6). $\blacksquare$

Example 2.3.7 Let $\lambda = (1^n)$ and fix

$$
t = \begin{array}{c} 1 \\ 2 \\ \vdots \\ n \end{array} \ . \tag{2.2}
$$

Thus

$$\kappa_t = \sum_{\sigma \in \mathcal{S}_n} (\operatorname{sgn}\sigma)\sigma,$$

and e_t is the signed sum of all $n!$ permutations regarded as tabloids. Now for any permutation π, Lemma 2.3.3 yields

$$e_{\pi t} = \pi e_t = \sum_{\sigma \in S_n} (\mathrm{sgn}\,\sigma)\pi\sigma\{t\}.$$

Replacing $\pi\sigma$ by τ,

$$e_{\pi t} = \sum_{\tau \in S_n} (\mathrm{sgn}\,\pi^{-1}\tau)\tau\{t\} = (\mathrm{sgn}\,\pi^{-1}) \sum_{\tau \in S_n} (\mathrm{sgn}\,\tau)\tau\{t\} = (\mathrm{sgn}\,\pi)e_t$$

because $\mathrm{sgn}\,\pi^{-1} = \mathrm{sgn}\,\pi$. Thus every polytabloid is a scalar multiple of e_t, where t is given by equation (2.2). So

$$S^{(1^n)} = \mathbb{C}\{e_t\}$$

with the action $\pi e_t = (\mathrm{sgn}\,\pi)e_t$. This is the sign representation. ∎

Example 2.3.8 If $\lambda = (n-1, 1)$, then we can use the module isomorphism of Example 2.1.8 to abuse notation and write $(n-1, 1)$-tabloids as

$$\{t\} = \frac{\overline{\begin{array}{ccc} \mathbf{i} & \cdots & \mathbf{j} \end{array}}}{\underline{\mathbf{k}}} \stackrel{\mathrm{def}}{=} \mathbf{k}.$$

This tabloid has $e_t = \mathbf{k} - \mathbf{i}$, and the span of all such vectors is easily seen to be

$$S^{(n-1,1)} = \{c_1\mathbf{1} + c_2\mathbf{2} + \cdots + c_n\mathbf{n} \;:\; c_1 + c_2 + \cdots + c_n = 0\}.$$

So $\dim S^{(n-1,1)} = n - 1$, and we can choose a basis for this module, e.g.,

$$\mathcal{B} = \{\mathbf{2} - \mathbf{1}, \mathbf{3} - \mathbf{1}, \ldots, \mathbf{n} - \mathbf{1}\}.$$

Computing the action of $\pi \in S_n$ on $\mathcal{B}$, we see that the corresponding character is one less than the number of fixedpoints of π. Thus $S^{(n-1,1)}$ is the module found at the end of Example 1.9.5. ∎

The reader should verify that when $n = 3$, the preceding three examples do give all the irreducible representations of S_3 (which were found in Chapter 1 by other means). So at least in that case we have fulfilled our aim.

2.4 The Submodule Theorem

It is time to show that the S^λ constitute a full set of irreducible S_n-modules. The crucial result will be the submodule theorem of James [Jam 76]. All the results and proofs of this section, up to and including this theorem, are true over an arbitrary field. The only change needed is to substitute a bilinear form for the inner product of equation (2.3). The fact that this replaces linearity by conjugate linearity in the second variable is not a problem, since we never need to carry a nonreal scalar from one side to the other.

Recall that $H^- = \sum_{\pi \in H}(\operatorname{sgn}\pi)\pi$ for any subset $H \subseteq S_n$. If $H = \{\pi\}$, then we write π^- for H^-. We will also need the unique inner product on M^λ for which

$$\langle\{t\}, \{s\}\rangle = \delta_{\{t\},\{s\}}. \tag{2.3}$$

Lemma 2.4.1 (Sign Lemma) *Let $H \leq S_n$ be a subgroup.*

1. *If $\pi \in H$, then*

$$\pi H^- = H^- \pi = (\operatorname{sgn}\pi)H^-.$$

   Otherwise put: $\pi^- H^- = H^-$.

2. *For any $\mathbf{u}, \mathbf{v} \in M^\lambda$,*

$$\langle H^- \mathbf{u}, \mathbf{v}\rangle = \langle \mathbf{u}, H^- \mathbf{v}\rangle.$$

3. *If the transposition $(b, c) \in H$, then we can factor*

$$H^- = k(\epsilon - (b, c)),$$

   where $k \in \mathbb{C}[S_n]$.

4. *If t is a tableau with b, c in the same row of t and $(b, c) \in H$, then*

$$H^-\{t\} = 0.$$

Proof. 1. This is just like the proof that $\pi e_t = (\operatorname{sgn}\pi)e_t$ in Example 2.3.7.

2. Using the fact that our form is S_n-invariant,

$$\langle H^- \mathbf{u}, \mathbf{v}\rangle = \sum_{\pi \in H} \langle (\operatorname{sgn}\pi)\pi\mathbf{u}, \mathbf{v}\rangle = \sum_{\pi \in H} \langle \mathbf{u}, (\operatorname{sgn}\pi)\pi^{-1}\mathbf{v}\rangle.$$

Replacing π by π^{-1} and noting that this does not affect the sign, we see that this last sum equals $\langle \mathbf{u}, H^- \mathbf{v}\rangle$.

3. Consider the subgroup $K = \{\epsilon, (b, c)\}$ of H. Then we can find a transversal and write $H = \biguplus_i k_i K$. But then $H^- = (\sum_i k_i^-)(\epsilon - (b, c))$, as desired.

4. By hypothesis, $(b, c)\{t\} = \{t\}$. Thus

$$H^-\{t\} = k(\epsilon - (b, c))\{t\} = k(\{t\} - \{t\}) = 0. \blacksquare$$

The sign lemma has a couple of useful corollaries.

Corollary 2.4.2 *Let $t = t^\lambda$ be a λ-tableau and $s = s^\mu$ be a μ-tableau, where $\lambda, \mu \vdash n$. If $\kappa_t\{s\} \neq 0$, then $\lambda \trianglerighteq \mu$. And if $\lambda = \mu$, then $\kappa_t\{s\} = \pm e_t$.*

Proof. Suppose b and c are two elements in the same row of s^μ. Then they cannot be in the same column of t^λ, for if so, then $\kappa_t = k(\epsilon - (b,c))$ and $\kappa_t\{s\} = 0$ by parts 3 and 4 in the preceding lemma. Thus the dominance lemma (Lemma 2.2.4) yields $\lambda \trianglerighteq \mu$.

If $\lambda = \mu$, then we must have $\{s\} = \pi\{t\}$ for some $\pi \in C_t$ by the same argument that established the dominance lemma. Using part 1 yields

$$\kappa_t\{s\} = \kappa_t\pi\{t\} = (\operatorname{sgn}\pi)\kappa_t\{t\} = \pm e_t. \quad \blacksquare$$

Corollary 2.4.3 *If* $\mathbf{u} \in M^\mu$ *and* $\operatorname{sh} t = \mu$, *then* $\kappa_t\mathbf{u}$ *is a multiple of* e_t.

Proof. We can write $\mathbf{u} = \sum_i c_i\{s_i\}$, where the s_i are μ-tableaux. By the previous corollary, $\kappa_t\mathbf{u} = \sum_i \pm c_i e_t$. $\quad \blacksquare$

We are now in a position to prove the submodule theorem.

Theorem 2.4.4 (Submodule Theorem [Jam 76]) *Let* U *be a submodule of* M^μ. *Then*

$$U \supseteq S^\mu \quad or \quad U \subseteq S^{\mu\perp}.$$

In particular, when the field is $\mathbb{C}$, *the* S^μ *are irreducible.*

Proof. Consider $\mathbf{u} \in U$ and a μ-tableau t. By the preceding corollary, we know that $\kappa_t\mathbf{u} = fe_t$ for some field element f. There are two cases, depending on which multiples can arise.

Suppose that there exits a $\mathbf{u}$ and a t with $f \neq 0$. Then since $\mathbf{u}$ is in the submodule U, we have $fe_t = \kappa_t\mathbf{u} \in U$. Thus $e_t \in U$ (since f is nonzero) and $S^\mu \subseteq U$ (since S^μ is cyclic).

On the other hand, suppose we always have $\kappa_t\mathbf{u} = \mathbf{0}$. We claim that this forces $U \subseteq S^{\mu\perp}$. Consider any $\mathbf{u} \in U$. Given an arbitrary μ-tableau t, we can apply part 2 of the sign lemma to obtain

$$\langle \mathbf{u}, e_t \rangle = \langle \mathbf{u}, \kappa_t\{t\}\rangle = \langle \kappa_t\mathbf{u}, \{t\}\rangle = \langle \mathbf{0}, \{t\}\rangle = \mathbf{0}.$$

Since the e_t span S^μ, we have $\mathbf{u} \in S^{\mu\perp}$, as claimed. $\quad \blacksquare$

It is only now that we will need our field to be the complexes (or any field of characteristic 0).

Proposition 2.4.5 *Suppose the field of scalars is* $\mathbb{C}$ *and* $\theta \in \operatorname{Hom}(S^\lambda, M^\mu)$ *is nonzero. Thus* $\lambda \trianglerighteq \mu$, *and if* $\lambda = \mu$, *then* θ *is multiplication by a scalar.*

Proof. Since $\theta \neq 0$, there is some basis vector e_t such that $\theta(e_t) \neq \mathbf{0}$. Because $\langle \cdot, \cdot \rangle$ is an inner product with complex scalars, $M^\lambda = S^\lambda \oplus S^{\lambda\perp}$. Thus we can extend θ to an element of $\operatorname{Hom}(M^\lambda, M^\mu)$ by setting $\theta(S^{\lambda\perp}) = \mathbf{0}$. So

$$\mathbf{0} \neq \theta(e_t) = \theta(\kappa_t\{t\}) = \kappa_t\theta(\{t\}) = \kappa_t(\sum_i c_i\{s_i\}),$$

where the s_i are μ-tableaux. By Corollary 2.4.2 we have $\lambda \trianglerighteq \mu$.

In the case $\lambda = \mu$, Corollary 2.4.3 yields $\theta(e_t) = ce_t$ for some constant c. So for any permutation π,

$$\theta(e_{\pi t}) = \theta(\pi e_t) = \pi\theta(e_t) = \pi(ce_t) = ce_{\pi t}.$$

Thus θ is multiplication by c. ∎

We can finally verify all our claims about the Specht modules.

Theorem 2.4.6 *The S^λ for $\lambda \vdash n$ form a complete list of irreducible S_n-modules over the complex field.*

Proof. The S^λ are irreducible by the submodule theorem and the fact that $S^\lambda \cap S^{\lambda\perp} = 0$ for the field $\mathbb{C}$.

Since we have the right number of modules for a full set, it suffices to show that they are pairwise inequivalent. But if $S^\lambda \cong S^\mu$, then there is a nonzero homomorphism $\theta \in \text{Hom}(S^\lambda, M^\mu)$, since $S^\mu \subseteq M^\mu$. Thus $\lambda \trianglerighteq \mu$ (Proposition 2.4.5). Similarly, $\mu \trianglerighteq \lambda$, so $\lambda = \mu$. ∎

Although the Specht modules are not necessarily irreducible over a field of characteristic p, p prime, the submodule theorem says that the quotient $S^\lambda/(S^\lambda \cap S^{\lambda\perp})$ is. These are the objects that play the role of S^λ in the theory of p-modular representations of S_n. See James [Jam 78] for further details.

Corollary 2.4.7 *The permutation modules decompose as*

$$M^\mu = \bigoplus_{\lambda \trianglerighteq \mu} m_{\lambda\mu} S^\lambda$$

with the diagonal multiplicity $m_{\mu\mu} = 1$.

Proof. This result follows from Proposition 2.4.5. If S^λ appears in M^μ with nonzero coefficient, then $\lambda \trianglerighteq \mu$. If $\lambda = \mu$, then we can also apply Proposition 1.7.10 to obtain

$$m_{\mu\mu} = \dim \text{Hom}(S^\mu, M^\mu) = 1. \blacksquare$$

The coefficients $m_{\lambda\mu}$ have a nice combinatorial interpretation, as we will see in Section 2.11.

2.5 Standard Tableaux and a Basis for S^λ

In general, the polytabloids that generate S^λ are not independent. It would be nice to have a subset forming a basis—e.g., for computing the matrices and characters of the representation. There is a very natural set of tableaux that can be used to index a basis.

Definition 2.5.1 A tableau t is *standard* if the rows and columns of t are increasing sequences. In this case we also say that the corresponding tabloid and polytabloid are standard. ∎

For example,

$$t = \begin{array}{ccc} 1 & 2 & 3 \\ 4 & 6 & \\ 5 & & \end{array}$$

is standard, but

$$t = \begin{array}{ccc} 1 & 2 & 3 \\ 5 & 4 & \\ 6 & & \end{array}$$

is not.

The next theorem is true over an arbitrary field.

Theorem 2.5.2 *The set*

$$\{e_t \; : \; t \text{ is a standard } \lambda\text{-tableau }\}$$

is a basis for S^λ.

We will spend the next two sections proving this theorem. First we will establish that the e_t are independent. As before, we will need a partial order, this time on tabloids.

It is convenient at this point to consider ordered partitions.

Definition 2.5.3 A *composition of n* is an ordered sequence of nonnegative integers

$$\lambda = (\lambda_1, \lambda_2, \ldots, \lambda_l)$$

such that $\sum_i \lambda_i = n$. The integers λ_i are called the *parts* of the composition. ∎

Note that there is no weakly decreasing condition on the parts in a composition. Thus $(1, 3, 2)$ and $(3, 2, 1)$ are both compositions of 6, but only the latter is a partition. The definitions of a Ferrers diagram and tableau are extended to compositions in the obvious way. (However, there are no standard λ-tableaux if λ is not a partition, since places to the right of or below the shape of λ are considered to be filled with an infinity symbol. Thus columns do not increase for a general composition.) The dominance order on compositions is defined exactly as in Definition 2.2.2, only now $\lambda_1, \ldots, \lambda_i$ and $\mu_1, \ldots, \mu_i$ are just the first i parts of their respective compositions, not necessarily the i largest.

Now suppose that $\{t\}$ is a tabloid with $\operatorname{sh} t = \lambda \vdash n$. For each index i, $1 \leq i \leq n$, let

$$\{t^i\} = \text{the tabloid formed by all elements} \leq i \text{ in } \{t\}$$

and

$$\lambda^i = \text{the composition which is the shape of } \{t^i\}$$

As an example, consider

$$\{t\} = \begin{array}{|cc|} \hline 2 & 4 \\ \hline 1 & 3 \\ \hline \end{array}.$$

Then

$$\{t^1\} = \begin{array}{|c|} \hline \emptyset \\ \hline 1 \\ \hline \end{array}, \qquad \{t^2\} = \begin{array}{|c|} \hline 2 \\ \hline 1 \\ \hline \end{array}, \qquad \{t^3\} = \begin{array}{|cc|} \hline 2 \\ \hline 1 & 3 \\ \hline \end{array}, \qquad \{t^4\} = \begin{array}{|cc|} \hline 2 & 4 \\ \hline 1 & 3 \\ \hline \end{array},$$

$$\lambda^1 = (0,1), \qquad \lambda^2 = (1,1), \qquad \lambda^3 = (1,2), \qquad \lambda^4 = (2,2).$$

The dominance order on tabloids is determined by the dominance ordering on the corresponding compositions.

Definition 2.5.4 Let $\{s\}$ and $\{t\}$ be two tabloids with composition sequences λ^i and μ^i, respectively. Then $\{s\}$ *dominates* $\{t\}$, written $\{s\} \trianglerighteq \{t\}$, if $\lambda^i \trianglerighteq \mu^i$ for all i. ∎

The Hasse diagram for this ordering of the $(2,2)$-tabloids is as follows:

$$
\begin{array}{|cc|} \hline 1 & 2 \\ \hline 3 & 4 \\ \hline \end{array}
$$
$$\uparrow$$
$$
\begin{array}{|cc|} \hline 1 & 3 \\ \hline 2 & 4 \\ \hline \end{array}
$$

$$
\begin{array}{|cc|} \hline 2 & 3 \\ \hline 1 & 4 \\ \hline \end{array}
\qquad\qquad
\begin{array}{|cc|} \hline 1 & 4 \\ \hline 2 & 3 \\ \hline \end{array}
$$

$$
\begin{array}{|cc|} \hline 2 & 4 \\ \hline 1 & 3 \\ \hline \end{array}
$$
$$\uparrow$$
$$
\begin{array}{|cc|} \hline 3 & 4 \\ \hline 1 & 2 \\ \hline \end{array}
$$

Just as for partitions, there is a dominance lemma for tabloids.

Lemma 2.5.5 (Dominance Lemma for Tabloids) *If* $k < l$ *and* k *appears in a lower row than* l *in* $\{t\}$, *then*

$$\{t\} \triangleleft (k,l)\{t\}.$$

Proof. Suppose that $\{t\}$ and $(k,l)\{t\}$ have composition sequences λ^i and μ^i, respectively. Then for $i < k$ or $i \geq l$ we have $\lambda^i = \mu^i$.

Now consider the case where $k \leq i < l$. If r and q are the rows of $\{t\}$ in which k and l appear, respectively, then

$$
\begin{aligned}
\lambda^i \;=\; & \mu^i \text{ with the } q\text{th part decreased by } 1 \\
& \text{and the } r\text{th part increased by } 1.
\end{aligned}
$$

Since $q < r$ by assumption, $\lambda^i \lhd \mu^i$. ∎

If $\mathbf{v} = \sum_i c_i \{t_i\} \in M^\mu$, then we say that $\{t_i\}$ *appears in* $\mathbf{v}$ if $c_i \neq 0$. The dominance lemma puts a restriction on which tableaux can appear in a standard polytabloid.

Corollary 2.5.6 *If t is standard and $\{s\}$ appears in e_t, then $\{t\} \unrhd \{s\}$.*

Proof. Let $s = \pi t$, where $\pi \in C_t$, so that $\{s\}$ appears in e_t. We induct on the number of *column inversions* in s, i.e., the number of pairs $k < l$ in the same column of s such that k is in a lower row than l. Given any such pair,

$$\{s\} \lhd (k, l)\{s\}$$

by Lemma 2.5.5. But $(k, l)\{s\}$ has fewer inversions than $\{s\}$, so, by induction, $(k, l)\{s\} \unlhd \{t\}$ and the result follows. ∎

The previous corollary says that $\{t\}$ is the maximum tabloid in e_t, by which we mean the following.

Definition 2.5.7 Let $(A, \leq)$ be a poset. Then an element $b \in A$ is the *maximum* if $b \geq c$ for all $c \in A$. An element b is a *maximal* element if there is no $c \in A$ with $c > b$. *Minimum* and *minimal* elements are defined analogously. ∎

Thus a maximum element is maximal, but the converse is not necessarily true. It is important to keep this distinction in mind in the next result.

Lemma 2.5.8 *Let $\mathbf{v}_1, \mathbf{v}_2, \ldots, \mathbf{v}_m$ be elements of M^μ. Suppose, for each $\mathbf{v}_i$, we can choose a tabloid $\{t_i\}$ appearing in $\mathbf{v}_i$ such that*

1. *$\{t_i\}$ is maximum in $\mathbf{v}_i$, and*

2. *the $\{t_i\}$ are all distinct.*

Then $\mathbf{v}_1, \mathbf{v}_2, \ldots, \mathbf{v}_m$ are independent.

Proof. Choose the labels such that $\{t_1\}$ is maximal among the $\{t_i\}$. Thus conditions 1 and 2 ensure that $\{t_1\}$ appears only in $\mathbf{v}_1$. (If $\{t_1\}$ occurs in $\mathbf{v}_i$, $i > 1$, then $\{t_1\} \lhd \{t_i\}$, contradicting the choice of $\{t_1\}$.) It follows that in any linear combination

$$c_1 \mathbf{v}_1 + c_2 \mathbf{v}_2 + \cdots + c_m \mathbf{v}_m = 0$$

we must have $c_1 = 0$ because there is no other way to cancel $\{t_1\}$. By induction on m, the rest of the coefficients must also be zero. ∎

The reader should note two things about this lemma. First of all, it is not sufficient only to have the $\{t_i\}$ maximal in $\mathbf{v}_i$; it is easy to construct counterexamples. Also, we have used no special properties of $\unrhd$ in the proof, so the result remains true for any partial order on tabloids.

We now have all the ingredients to prove independence of the standard basis.

Proposition 2.5.9 *The set*

$$\{e_t \; : \; t \text{ is a standard } \lambda\text{-tableau}\}$$

is independent.

Proof. By Corollary 2.5.6, $\{t\}$ is maximum in e_t, and by hypothesis they are all distinct. Thus Lemma 2.5.8 applies. ∎

2.6 Garnir Elements

To show that the standard polytabloids of shape μ span S^μ, we use a procedure known as a *straightening algorithm*. The basic idea is this. Pick an arbitrary tableau t. We must show that e_t is a linear combination of standard polytabloids. We may as well assume that the columns of t are increasing, since if not, there is $\sigma \in C_t$ such that $s = \sigma t$ has increasing columns. So

$$e_s = \sigma e_t = (\text{sgn } \sigma) e_t$$

by Lemmas 2.3.3 (part 4) and 2.4.1 (part 1). Thus e_t is is a linear combination of polytabloids whenever e_s is.

Now suppose we can find permutations π such that

1. in each tableau πt, a certain *row descent* of t (pair of adjacent, out-of-order elements in a row) has been eliminated, and

2. the group algebra element $g = \epsilon + \sum_\pi (\text{sgn } \pi)\pi$ satisfies $g e_t = 0$.

Then

$$e_t = -\sum_\pi e_{\pi t}.$$

So we have expressed e_t in terms of polytabloids that are closer to being standard, and induction applies to obtain e_t as a linear combination of polytabloids.

The elements of the group algebra that accomplish this task are the Garnir elements.

Definition 2.6.1 Let A and B be two disjoint sets of positive integers and choose permutations π such that

$$\mathcal{S}_{A \cup B} = \biguplus_\pi \pi (\mathcal{S}_A \times \mathcal{S}_B).$$

Then a corresponding *Garnir element* is

$$g_{A,B} = \sum_\pi (\text{sgn } \pi)\pi. \quad \blacksquare$$

Although $g_{A,B}$ depends on the transversal and not just on A and B, we will standardize the choice of the π's in a moment. Perhaps the simplest way to obtain a transversal is as follows. The group $S_{A \cup B}$ acts on all ordered pairs (A', B') such that $|A'| = |A|$, $|B'| = |B|$, and $A' \uplus B' = A \uplus B$ in the obvious manner. If for each possible (A', B') we take $\pi \in S_{A \cup B}$ such that

$$\pi(A, B) = (A', B'),$$

then the collection of such permutations forms a transversal. For example, suppose $A = \{5, 6\}$ and $B = \{2, 4\}$. Then the corresponding pairs of sets (set brackets and commas having been eliminated for readability) and possible permutations are

$$(A', B') : (56, 24)\,,\,(46, 25)\,,\,(26, 45)\,,\,(45, 26)\,,\,(25, 46)\,,\quad(24, 56).$$
$$g_{A,B} = \quad\epsilon\quad-\quad(4, 5)\;-\;(2, 5)\;-\;(4, 6)\;-\;(2, 6)\;+\;(2, 5)(4, 6).$$

It should be emphasized that for any given pair (A', B'), there are many different choices for the permutation π sending (A, B) to that pair.

The Garnir element associated with a tableau t is used to eliminate a descent $t_{i,j} > t_{i,j+1}$.

Definition 2.6.2 Let t be a tableau and let A and B be subsets of the jth and $(j+1)$st columns of t, respectively. The *Garnir element associated with t* (and A, B) is $g_{A,B} = \sum_{\pi}(\operatorname{sgn}\pi)\pi$, where the π have been chosen so that the elements of $A \cup B$ are increasing down the columns of πt. ∎

In practice, we always take A (respectively, B) to be all elements below $t_{i,j}$ (respectively, above $t_{i,j+1}$), as in the diagram

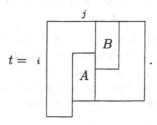

If we use the tableau

$$t = \begin{array}{ll} 1\ 2\ 3 \\ 5\ 4 \\ 6 \end{array}$$

with the descent $5 > 4$, then the sets A and B are the same as in the previous example. Each (A', B') has a corresponding t' that determines a permutation in $g_{A,B}$:

$$t' : \begin{array}{l} 1\ 2\ 3 \\ 5\ 4 \\ 6 \end{array}, \quad \begin{array}{l} 1\ 2\ 3 \\ 4\ 5 \\ 6 \end{array}, \quad \begin{array}{l} 1\ 4\ 3 \\ 2\ 5 \\ 6 \end{array}, \quad \begin{array}{l} 1\ 2\ 3 \\ 4\ 6 \\ 5 \end{array}, \quad \begin{array}{l} 1\ 4\ 3 \\ 2\ 6 \\ 5 \end{array}, \quad \begin{array}{l} 1\ 5\ 3 \\ 2\ 6 \\ 4 \end{array},$$
$$g_{A,B} = \quad\epsilon\;-\quad(4, 5)\;+\;(2, 4, 5)\;+\;(4, 6, 5)\;-\;(2, 4, 6, 5)\;+\;(2, 5)(4, 6).$$

The reader can verify that $g_{A,B}e_t = \mathbf{0}$, so that

$$e_t = e_{t_2} - e_{t_3} - e_{t_4} + e_{t_5} - e_{t_6},$$

where $t_2, \ldots, t_6$ are the second through sixth tableaux in the preceding list. Note that none of these arrays have the descent found in the second row of t.

Proposition 2.6.3 *Let t, A, and B, be as in the definition of a Garnir element. If $|A \cup B|$ is greater than the number of elements in column j of t, then $g_{A,B}e_t = \mathbf{0}$.*

Proof. First, we claim that

$$S^-_{A \cup B}e_t = \mathbf{0}. \tag{2.4}$$

Consider any $\sigma \in C_t$. By the hypothesis, there must be $a, b \in A \cup B$ such that a and b are in the same row of σt. But then $(a, b) \in S_{A \cup B}$ and $S^-_{A \cup B}\{\sigma t\} = \mathbf{0}$ by part 4 of the sign lemma (Lemma 2.4.1). Since this is true of every σ appearing in κ_t, the claim follows.

Now $S_{A \cup B} = \biguplus_\pi \pi(S_A \times S_B)$, so $S^-_{A \cup B} = g_{A,B}(S_A \times S_B)^-$. Substituting this into equation (2.4) yields

$$g_{A,B}(S_A \times S_B)^- e_t = \mathbf{0}, \tag{2.5}$$

and we need worry only about the contribution of $(S_A \times S_B)^-$. But we have $S_A \times S_B \subseteq C_t$. So if $\sigma \in S_A \times S_B$, then, by part 1 of the sign lemma,

$$\sigma^- e_t = \sigma^- C_t^-\{t\} = C_t^-\{t\} = e_t.$$

Thus $(S_A \times S_B)^- e_t = |S_A \times S_B|e_t$, and dividing equation (2.5) by this cardinality yields the proposition. ∎

The reader may have noticed that when we eliminated the descent in row 2 of the preceding example, we introduced descents in some other places— e.g., in row 1 of t_3. Thus we need some measure of standardness that makes $t_2, \ldots, t_6$ closer to being standard than t. This is supplied by yet another partial order. Given t, consider its *column equivalence class*, or *column tabloid*,

$$[t] \stackrel{\text{def}}{=} C_t t,$$

i.e., the set of all tableaux obtained by rearranging elements within columns of t. We use vertical lines to denote the column tabloid, as in

$$\left| \begin{array}{c|c} 1 & 2 \\ 3 & \end{array} \right| = \left\{ \begin{array}{ccc} 1 & 2 & 3 & 2 \\ 3 & & 1 & \end{array} \right\}.$$

Replacing "row" by "column" in the definition of dominance for tabloids, we obtain a definition of column dominance for which we use the same symbol as for rows (the difference in the types of brackets used for the classes makes the necessary distinction).

Our proof that the standard polytabloids span S^λ follows Peel [Pee 75].

Theorem 2.6.4 *The set*

$$\{e_t \; : \; t \text{ is a standard } \lambda\text{-tableau}\} \tag{2.6}$$

spans S^λ.

First note that if e_t is in the span of the set (2.6), then so is e_s for any $s \in [t]$, by the remarks at the beginning of this section. Thus we may always take t to have increasing columns.

The poset of column tabloids has a maximum element $[t_0]$, where t_0 is obtained by numbering the cells of each column consecutively from top to bottom, starting with the leftmost column and working right. Since t_0 is standard, we are done for this equivalence class.

Now pick any tableau t. By induction, we may assume that every tableau s with $[s] \rhd [t]$ is in the span of (2.6). If t is standard, then we are done. If not, then there must be a descent in some row i (since columns increase). Let the columns involved be the jth and $(j+1)$st with entries $a_1 < a_2 < \cdots < a_p$ and $b_1 < b_2 < \cdots < b_q$, respectively. Thus we have the following situation in t:

$$
\begin{array}{ccc}
a_1 & & b_1 \\
 & & \wedge \\
a_2 & & b_2 \\
 & & \wedge \\
\vdots & & \vdots \\
 & & \wedge \\
a_i & > & b_i \\
\wedge & & \\
\vdots & & \vdots \\
\wedge & & b_q \\
a_p & &
\end{array}
$$

Take $A = \{a_i, \ldots, a_p\}$ and $B = \{b_1, \ldots, b_i\}$ with associated Garnir element $g_{A,B} = \sum_\pi (\operatorname{sgn} \pi)\pi$. By Proposition 2.6.3 we have $g_{A,B} e_t = 0$, so that

$$e_t = -\sum_{\pi \neq \epsilon} (\operatorname{sgn} \pi) e_{\pi t}. \tag{2.7}$$

But $b_1 < \cdots < b_i < a_i < \cdots < a_p$ implies that $[\pi t] \unrhd [t]$ for $\pi \neq \epsilon$ by the column analogue of the dominance lemma for tabloids (Lemma 2.5.5). Thus all terms on the right side of (2.7), and hence e_t itself, are in the span of the standard polytabloids. ∎

To summarize our results, let

$$f^\lambda = \text{the number of standard } \lambda\text{-tableaux}.$$

Then the following is true over any base field.

Theorem 2.6.5 *For any partition* λ:

1. $\{e_t \ : \ t \text{ is a standard } \lambda\text{-tableau}\}$ *is a basis for* S^λ,

2. $\dim S^\lambda = f^\lambda$, *and*

3. $\sum_{\lambda \vdash n} (f^\lambda)^2 = n!$.

Proof. The first two parts are immediate. The third follows from the fact (Proposition 1.10.1) that for any group G,

$$\sum_V (\dim V)^2 = |G|,$$

where the sum is over all irreducible G-modules. ∎

2.7 Young's Natural Representation

The matrices for the module S^λ in the standard basis form what is known as *Young's natural representation*. In this section we illustrate how these arrays are obtained.

Since S_n is generated by the transpositions $(k, k+1)$ for $k = 1, \ldots, n-1$, one need compute only the matrices corresponding to these group elements. If t is a standard tableau, we get the tth column of the matrix for $\pi \in S_n$ by expressing $\pi e_t = e_{\pi t}$ as a sum of standard polytabloids. When $\pi = (k, k+1)$, there are three cases.

1. If k and $k+1$ are in the same column of t, then $(k, k+1) \in C_t$ and

   $$(k, k+1)e_t = -e_t.$$

2. If k and $k+1$ are in the same row of t, then $(k, k+1)t$ has a descent in that row. Applying the appropriate Garnir element, we obtain

   $$(k, k+1)e_t = e_t \pm \text{ other polytabloids } e_{t'} \text{ such that } [t'] \rhd [t].$$

3. If k and $k+1$ are not in the same row or column of t, then the tableau $t' = (k, k+1)t$ is standard and

   $$(k, k+1)e_t = e_{t'}.$$

The e_t in the sum for case 2 comes from the term of equation (2.7) corresponding to $\pi = (k, k+1)$. Although we do not have an explicit expression for the rest of the terms, repeated application of 1 through 3 will compute them (by reverse induction on the column dominance ordering). This is the straightening algorithm mentioned at the beginning of Section 2.6.

By way of example, let us compute the matrices for the representation of S_3 corresponding to $\lambda = (2,1)$. The two standard tableaux of shape λ are

$$t_1 = \begin{array}{ll} 1 & 3 \\ 2 \end{array} \quad \text{and} \quad t_2 = \begin{array}{ll} 1 & 2 \\ 3 \end{array} .$$

Note that we have chosen our indices so that $[t_1] \triangleright [t_2]$. This makes the computations in case 2 easier.

For the transposition $(1,2)$ we have

$$(1,2)e_{t_1} = \frac{\boxed{\begin{array}{ll} 2 & 3 \\ 1 \end{array}}}{} - \frac{\boxed{\begin{array}{ll} 1 & 3 \\ 2 \end{array}}}{} = -e_{t_1}$$

as predicted by case 1. Since

$$(1,2)t_2 = \begin{array}{ll} 2 & 1 \\ 3 \end{array}$$

has a descent in row 1, we must find a Garnir element. Taking $A = \{2,3\}$ and $B = \{1\}$ gives tableaux

$$\begin{array}{ll} 2 & 1 \\ 3 \end{array} = (1,2)t_2, \quad \begin{array}{ll} 1 & 2 \\ 3 \end{array} = t_2, \quad \text{and} \quad \begin{array}{ll} 1 & 3 \\ 2 \end{array} = t_1$$

with element

$$g_{A,B} = \epsilon - (1,2) + (1,3,2).$$

Thus

$$e_{(1,2)t_2} - e_{t_2} + e_{t_1} = 0,$$

or

$$(1,2)e_{t_2} = e_{t_2} - e_{t_1}$$

(This result can be obtained more easily by merely noting that

$$(1,2)e_{t_2} = \frac{\boxed{\begin{array}{ll} 1 & 2 \\ 3 \end{array}}}{} - \frac{\boxed{\begin{array}{ll} 3 & 1 \\ 2 \end{array}}}{}$$

and expressing this as a linear combination of the e_{t_i}.) Putting everything together, we obtain the matrix

$$X((1,2)) = \begin{pmatrix} -1 & -1 \\ 0 & 1 \end{pmatrix} .$$

The transposition $(2,3)$ is even easier to deal with, since case 3 always applies, yielding

$$(2,3)t_1 = t_2$$

and

$$(2,3)t_2 = t_1.$$

Thus

$$X((2,3)) = \begin{pmatrix} 0 & 1 \\ 1 & 0 \end{pmatrix} .$$

We have already seen these two matrices in Example 1.9.5. Since the adjacent transpositions generate S_n, all other matrices must agree as well.

2.8 The Branching Rule

It is natural to ask what happens when we restrict or induce an irreducible representation S^λ of $\mathcal{S}_n$ to $\mathcal{S}_{n-1}$ or $\mathcal{S}_{n+1}$, respectively. The branching theorem gives a simple answer to that question.

Intuitively, these two operations correspond to either removing or adding a node to the Ferrers diagram for λ.

Definition 2.8.1 If λ is a diagram, then an *inner corner of* λ is a node $(i, j) \in \lambda$ whose removal leaves the Ferrers diagram of a partition. Any partition obtained by such a removal is denoted by λ^-. An *outer corner of* λ is a node $(i, j) \notin \lambda$ whose addition produces the Ferrers diagram of a partition. Any partition obtained by such an addition is denoted by λ^+. ∎

Note that the inner corners of λ are exactly those nodes at the end of a row and column of λ. For example, if $\lambda = (5, 4, 4, 2)$, then the inner corners are enlarged and the outer corners marked with open circles in the following diagram:

So, after removal, we could have

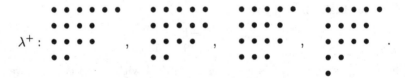

whereas after addition, the possibilities are

$\lambda^+:$

These are exactly the partitions that occur in restriction and induction. In particular,

$$S^{(5,4,4,2)}\!\downarrow_{\mathcal{S}_{14}} \cong S^{(4,4,4,2)} \oplus S^{(5,4,3,2)} \oplus S^{(5,4,4,1)}$$

and

$$S^{(5,4,4,2)}\!\uparrow^{\mathcal{S}_{16}} \cong S^{(6,4,4,2)} \oplus S^{(5,5,4,2)} \oplus S^{(5,4,4,3)} \oplus S^{(5,4,4,2,1)}.$$

Before proving the branching rule itself, we need a result about dimensions.

Lemma 2.8.2 *We have*

$$f^\lambda = \sum_{\lambda^-} f^{\lambda^-}.$$

Proof. Every standard tableau of shape $\lambda \vdash n$ consists of n in some inner corner together with a standard tableau of shape $\lambda^- \vdash n - 1$. The result follows. ∎

Theorem 2.8.3 (Branching Rule) *If $\lambda \vdash n$, then*

1. $S^\lambda \!\downarrow_{S_{n-1}} \cong \bigoplus_{\lambda^-} S^{\lambda^-}$, *and*

2. $S^\lambda \!\uparrow^{S_{n+1}} \cong \bigoplus_{\lambda^+} S^{\lambda^+}$.

Proof. [Pee 75] 1. Let the inner corners of λ appear in rows $r_1 < r_2 < \cdots < r_k$. For each i, let λ^i denote the partition λ^- obtained by removing the corner cell in row r_i. In addition, if n is at the end of row r_i of tableau t (respectively, in row r_i of tabloid $\{t\}$), then t^i (respectively, $\{t^i\}$) will be the array obtained by removing the n.

Now given any group G with module V and submodule W, it is easy to see that

$$V \cong W \oplus (V/W),$$

where V/W is the quotient space. (See Exercise 7 in Chapter 1.) Thus it suffices to find a chain of subspaces

$$\{0\} = V^{(0)} \subset V^{(1)} \subset V^{(2)} \subset \cdots \subset V^{(k)} = S^\lambda$$

such that $V^{(i)}/V^{(i-1)} \cong S^{\lambda^i}$ as S_{n-1}-modules for $1 \le i \le k$. Let $V^{(i)}$ be the vector space spanned by the standard polytabloids e_t, where n appears in t at the end of one of rows r_1 through r_i. We show that the $V^{(i)}$ are our desired modules as follows.

Define maps $\theta_i : M^\lambda \to M^{\lambda^i}$ by linearly extending

$$\{t\} \overset{\theta_i}{\to} \begin{cases} \{t_i\} & \text{if } n \text{ is in row } r_i \text{ of } \{t\}, \\ 0 & \text{otherwise.} \end{cases}$$

The reader can quickly verify that θ_i is an S_{n-1}-homomorphism. Furthermore, for standard t we have

$$e_t \overset{\theta_i}{\to} \begin{cases} e_{t^i} & \text{if } n \text{ is in row } r_i \text{ of } t, \\ 0 & \text{if } n \text{ is in row } r_j \text{ of } t, \text{ where } j < i. \end{cases} \tag{2.8}$$

This is because any tabloid appearing in e_t, t standard, has n in the same row or higher than in t.

Since the standard polytabloids form a basis for the corresponding Specht module, the two parts of (2.8) show that

$$\theta_i V^{(i)} = S^{\lambda^i} \tag{2.9}$$

and
$$V^{(i-1)} \subseteq \ker \theta_i. \qquad (2.10)$$

From equation (2.10), we can construct the chain
$$\{0\} = V^{(0)} \subseteq V^{(1)} \cap \ker \theta_1 \subseteq V^{(1)} \subseteq V^{(2)} \cap \ker \theta_2 \subseteq V^{(2)} \subseteq \cdots \subseteq V^{(k)} = S^\lambda. \qquad (2.11)$$

But from equation (2.9)
$$\dim \frac{V^{(i)}}{V^{(i)} \cap \ker \theta_i} = \dim \theta_i V^{(i)} = f^{\lambda^i}.$$

By the preceding lemma, the dimensions of these quotients add up to $\dim S^\lambda$. Since this leaves no space to insert extra modules, the chain (2.11) must have equality for the first, third, etc. containments. Furthermore,
$$\frac{V^{(i)}}{V^{(i-1)}} \cong \frac{V^{(i)}}{V^{(i)} \cap \ker \theta_i} \cong S^{\lambda^i}$$

as desired.

2. We will show that this part follows from the first by Frobenius reciprocity (Theorem 1.12.6). In fact, parts 1 and 2 can be shown to be equivalent by the same method.

Let χ^λ be the character of S^λ. If $S^\lambda \!\uparrow^{S_{n+1}} \cong \oplus_{\mu \vdash n+1} m_\mu S^\mu$, then by taking characters, $\chi^\lambda \!\uparrow^{S_{n+1}} \cong \sum_{\mu \vdash n+1} m_\mu \chi^\mu$. The multiplicities are given by

$$
\begin{aligned}
m_\mu &= \langle \chi^\lambda \!\uparrow^{S_{n+1}}, \chi^\mu \rangle && \text{(Corollary 1.9.4, part 2)} \\
&= \langle \chi^\lambda, \chi^\mu \!\downarrow_{S_n} \rangle && \text{(Frobenius reciprocity)} \\
&= \langle \chi^\lambda, \sum_{\mu^-} \chi^{\mu^-} \rangle && \text{(branching rule, part 1)} \\
&= \begin{cases} 1 & \text{if } \lambda = \mu^-, \\ 0 & \text{otherwise} \end{cases} && \text{(Corollary 1.9.4, part 4)} \\
&= \begin{cases} 1 & \text{if } \mu = \lambda^+, \\ 0 & \text{otherwise.} \end{cases} && \text{(definition of } \mu^- \text{ and } \lambda^+\text{)}
\end{aligned}
$$

This finishes the proof. ∎

2.9   The Decomposition of M^μ

We would like to know the multiplicity $m_{\lambda\mu}$ of the Specht module S^λ in M^μ. In fact, we can give a nice combinatorial description of these numbers in terms of tableaux that allow repetition of entries.

Definition 2.9.1 A *generalized Young tableau of shape* λ, is an array T obtained by replacing the nodes of λ with positive integers, repetitions allowed. The *type* or *content of* T is the composition $\mu = (\mu_1, \mu_2, \ldots, \mu_m)$, where μ_i equals the number of i's in T. Let
$$\mathcal{T}_{\lambda\mu} = \{T : T \text{ has shape } \lambda \text{ and content } \mu\}. \ \blacksquare$$

Note that capital letters will be used for generalized tableaux. One such array is

$$T = \begin{array}{ccc} 4 & 1 & 4 \\ 1 & 3 & \end{array}$$

of shape $(3, 2)$ and content $(2, 0, 1, 2)$.

We will show that $\mathbb{C}[\mathcal{T}_{\lambda\mu}]$ is really a new copy of M^μ. For the rest of this section and the next:

fix a tableau t of shape λ and content (1^n).

In all our examples we will use $\lambda = (3, 2)$ and

$$t = \begin{array}{ccc} 1 & 2 & 3 \\ 4 & 5 & \end{array} .$$

If T is any λ-tableau, then let $T(i)$ denote the element of T in the same position as the i in the fixed tableau t. With t and T as before,

$$T = \begin{array}{ccc} T(1) & T(2) & T(3) \\ T(4) & T(5) & \end{array} , \tag{2.12}$$

so that $T(1) = T(3) = 4$, $T(2) = T(4) = 1$, and $T(5) = 3$.

Now given any tabloid $\{s\}$ of shape μ, produce a tableau $T \in \mathcal{T}_{\lambda\mu}$ by letting

$T(i) =$ the number of the row in which i appears in $\{s\}$.

For example, suppose $\mu = (2, 2, 1)$ and

$$\{s\} = \begin{array}{cc} \overline{2 \quad 3} \\ \overline{1 \quad 5} \\ \overline{4} \end{array} .$$

Then the 2 and 3 are in row one of $\{s\}$, so the 2 and 3 of t get replaced by ones, etc., to obtain

$$T = \begin{array}{ccc} 2 & 1 & 1 \\ 3 & 2 & \end{array} .$$

Note that the shape of $\{s\}$ becomes the content of T, as desired. Also, it is clear that the map $\{s\} \overset{\theta}{\to} T$ is a bijection between bases for M^μ and $\mathbb{C}[\mathcal{T}_{\lambda\mu}]$, so they are isomorphic as vector spaces.

We now need to define an action of $\mathcal{S}_n$ on generalized tableaux so that $M^\mu \cong \mathbb{C}[\mathcal{T}_{\lambda\mu}]$ as modules. If $\pi \in \mathcal{S}_n$ and $T \in \mathcal{T}_{\lambda\mu}$, then we let πT be the tableau such that

$$(\pi T)(i) \overset{\text{def}}{=} T(\pi^{-1}i).$$

To illustrate with our canonical tableau (2.12):

$$(1, 2, 4)T = \begin{array}{ccc} T(4) & T(1) & T(3) \\ T(2) & T(5) & \end{array} .$$

In particular,

$$(1,2,4) \begin{array}{ccc} 2 & 1 & 1 \\ 3 & 2 & \end{array} = \begin{array}{ccc} 3 & 2 & 1 \\ 1 & 2 & \end{array} .$$

Note that although $\pi \in S_n$ acts on the elements of $\{s\}$ (replacing i by πi), it acts on the places in T (moving $T(i)$ to the position of $T(\pi i)$). To see why this is the correct definition for making θ into a module isomorphism, consider $\theta(\{s\}) = T$ and $\pi \in S_n$. Thus we want $\theta(\pi\{s\}) = \pi T$. But this forces us to set

$$\begin{aligned} (\pi T)(i) &= \text{row number of } i \text{ in } \pi\{s\} \\ &= \text{row number of } \pi^{-1}i \text{ in } \{s\} \\ &= T(\pi^{-1}i). \end{aligned}$$

We have proved, by fiat, the following.

Proposition 2.9.2 *For any given partition λ, the modules M^μ and $\mathbb{C}[T_{\lambda\mu}]$ are isomorphic.* ∎

Recall (Proposition 1.7.10) that the multiplicity of S^λ in M^μ is given by $\dim \mathrm{Hom}(S^\lambda, M^\mu)$. We will first construct certain homomorphisms from M^λ to M^μ using our description of the latter in terms of generalized tableaux and then restrict to S^λ. The row and column equivalence classes of a generalized tableau T, denoted by $\{T\}$ and $[T]$, respectively, are defined in the obvious way. Let $\{t\} \in M^\lambda$ be the tabloid associated with our fixed tableau.

Definition 2.9.3 For each $T \in T_{\lambda\mu}$, the *homomorphism corresponding to T* is the map $\theta_T \in \mathrm{Hom}(M^\lambda, M^\mu)$ given by

$$\{t\} \overset{\theta_T}{\to} \sum_{S \in \{T\}} S$$

and extension using the cyclicity of M^λ. Note that θ_T is actually a homomorphism into $\mathbb{C}T_{\lambda\mu}$, but that should cause no problems in view of the previous proposition. ∎

Extension by cyclicity means that, since every element of M^λ is of the form $g\{t\}$ for some $g \in \mathbb{C}[S_n]$, we must have $\theta_T(g\{t\}) = g \sum_{S \in \{T\}} S$ (that θ_T respects the group action and linearity). For example, if

$$T = \begin{array}{ccc} 2 & 1 & 1 \\ 3 & 2 & \end{array} .$$

then

$$\theta_T\{t\} = \begin{array}{ccc} 2 & 1 & 1 \\ 3 & 2 & \end{array} + \begin{array}{ccc} 1 & 2 & 1 \\ 3 & 2 & \end{array} + \begin{array}{ccc} 1 & 1 & 2 \\ 3 & 2 & \end{array} + \begin{array}{ccc} 2 & 1 & 1 \\ 2 & 3 & \end{array} + \begin{array}{ccc} 1 & 2 & 1 \\ 2 & 3 & \end{array} + \begin{array}{ccc} 1 & 1 & 2 \\ 2 & 3 & \end{array}$$

and

$$\theta_T(1,2,4)\{t\} = \begin{array}{ccc} 3 & 2 & 1 \\ 1 & 2 & \end{array} + \begin{array}{ccc} 3 & 1 & 1 \\ 2 & 2 & \end{array} + \begin{array}{ccc} 3 & 1 & 2 \\ 1 & 2 & \end{array} + \begin{array}{ccc} 2 & 2 & 1 \\ 1 & 3 & \end{array} + \begin{array}{ccc} 2 & 1 & 1 \\ 2 & 3 & \end{array} + \begin{array}{ccc} 2 & 1 & 2 \\ 1 & 3 & \end{array} .$$

Now we obtain elements of $\text{Hom}(S^\lambda, M^\mu)$ by letting

$$\bar{\theta}_T = \text{ the restriction of } \theta_T \text{ to } S^\lambda.$$

If t is our fixed tableau, then

$$\bar{\theta}_T(e_t) = \bar{\theta}_T(\kappa_t\{t\}) = \kappa_t(\theta_T\{t\}) = \kappa_t\left(\sum_{S \in \{T\}} S\right).$$

This last expression could turn out to be zero (thus forcing $\bar{\theta}_T$ to be the zero map by the cyclicity S^λ) because of the following.

Proposition 2.9.4 *If t is the fixed λ-tableau and $T \in \mathcal{T}_{\lambda\mu}$, then $\kappa_t T = 0$ if and only if T has two equal elements in some column.*

Proof. If $\kappa_t T = 0$, then

$$T + \sum_{\substack{\pi \in C_t \\ \pi \neq \epsilon}} (\text{sgn } \pi)\pi T = 0.$$

So we must have $T = \pi T$ for some $\pi \in C_t$ with $\text{sgn } \pi = -1$. But then the elements corresponding to any nontrivial cycle of π are all equal and in the same column.

Now suppose that $T(i) = T(j)$ are in the same column of T. Then

$$(\epsilon - (i,j))T = 0.$$

But $\epsilon - (i,j)$ is a factor of κ_t by part 3 of the sign lemma (Lemma 2.4.1), so $\kappa_t T = 0$. ∎

In light of the previous proposition, we can eliminate possibly trivial $\bar{\theta}_T$ from consideration by concentrating on the analogue of standard tableaux for arrays with repetitions.

Definition 2.9.5 A generalized tableau is *semistandard* if its rows weakly increase and its columns strictly increase. We let $\mathcal{T}^0_{\lambda\mu}$ denote the set of semi-standard λ-tableaux of type μ. ∎

The tableau

$$T = \begin{matrix} 1 & 1 & 2 \\ 2 & 3 & \end{matrix}$$

is semistandard, whereas

$$T = \begin{matrix} 2 & 1 & 1 \\ 3 & 2 & \end{matrix}$$

is not. The homomorphisms corresponding to semistandard tableaux are the ones we have been looking for. Specifically, we will show that they form a basis for $\text{Hom}(S^\lambda, M^\mu)$.

2.10 The Semistandard Basis for $\text{Hom}(S^\lambda, M^\mu)$

This section is devoted to proving the following theorem.

Theorem 2.10.1 *The set*

$$\{\bar{\theta}_T \; : \; T \in \mathcal{T}^0_{\lambda\mu}\}$$

is a basis for $\text{Hom}(S^\lambda, M^\mu)$.

In many ways the proof will parallel the one given in Sections 2.5 and 2.6 to show that the standard polytabloids form a basis for S^λ.

As usual, we need appropriate partial orders. The dominance and column dominance orderings for generalized tableaux are defined in exactly the same way as for tableaux without repetitions (see Definition 2.5.4). For example, if

$$[S] = \begin{array}{|c|c|c|} \hline 2 & 1 & 1 \\ \hline 3 & 2 \\ \cline{1-2} \end{array} \quad \text{and} \quad [T] = \begin{array}{|c|c|c|} \hline 1 & 1 & 2 \\ \hline 2 & 3 \\ \cline{1-2} \end{array},$$

then $[S]$ corresponds to the sequence of compositions

$$\lambda^1 = (0,1,1), \quad \lambda^2 = (1,2,1), \quad \lambda^3 = (2,2,1),$$

whereas $[T]$ has

$$\mu^1 = (1,1,0), \quad \mu^2 = (2,1,1), \quad \mu^3 = (2,2,1).$$

Since $\lambda^i \trianglelefteq \mu^i$ for all i, $[S] \trianglelefteq [T]$. The dominance lemma for tabloids (Lemma 2.5.5) and its corollary (Corollary 2.5.6) have analogues in this setting. Their proofs, being similar, are omitted.

Lemma 2.10.2 (Dominance Lemma for Generalized Tableaux) *Let k be in a column to the left of l in T with $k < l$. Then*

$$[T] \rhd [S],$$

where S is the tableau obtained by interchanging k and l in T. ∎

Corollary 2.10.3 *If T is semistandard and $S \in \{T\}$ is different from T, then*

$$[T] \rhd [S]. \blacksquare$$

Thus, if T is semistandard, then $[T]$ is the largest equivalence class to appear in $\theta_T\{t\}$.

Before proving independence of the $\bar{\theta}_T$, we must cite some general facts about vector spaces. Let V be a vector space and pick out a fixed basis $\mathcal{B} = \{b_1, b_2, \ldots, b_n\}$. If $\mathbf{v} \in V$, then we say that b_i *appears in* $\mathbf{v}$ if $\mathbf{v} = \sum_i c_i b_i$ with $c_i \neq 0$. Suppose that V is endowed with an equivalence relation whose equivalence classes are denoted by $[\mathbf{v}]$, and with a partial order on these classes. We can generalize Lemma 2.5.8 as follows.

Lemma 2.10.4 *Let V and $\mathcal{B}$ be as before and consider a set of vectors $\mathbf{v}_1, \mathbf{v}_2, \ldots, \mathbf{v}_m \in V$. Suppose that, for all i, there exists $\mathbf{b}_i \in \mathcal{B}$ appearing in $\mathbf{v}_i$ such that*

1. *$[\mathbf{b}_i] \trianglerighteq [\mathbf{b}]$ for every $\mathbf{b} \neq \mathbf{b}_i$ appearing in $\mathbf{v}_i$, and*

2. *the $[\mathbf{b}_i]$ are all distinct.*

Then the $\mathbf{v}_i$ are linearly independent. ∎

We also need a simple lemma about independence of linear transformations.

Lemma 2.10.5 *Let V and W be vector spaces and let $\theta_1, \theta_2, \ldots, \theta_m$ be linear maps from V to W. If there exists a $\mathbf{v} \in V$ such that $\theta_1(\mathbf{v}), \theta_2(\mathbf{v}), \ldots, \theta_m(\mathbf{v})$ are independent in W, then the θ_i are independent as linear transformations.*

Proof. Suppose not. Then there are constants c_i, not all zero, such that $\sum_i c_i \theta_i$ is the zero map. But then $\sum_i c_i \theta_i(\mathbf{v}) = \mathbf{0}$ for all $\mathbf{v} \in V$, a contradiction to the hypothesis of the lemma. ∎

Proposition 2.10.6 *The set*

$$\{\bar{\theta}_T \; : \; T \in \mathcal{T}^0_{\lambda\mu}\}$$

is independent.

Proof. Let $T_1, T_2, \ldots, T_m$ be the elements of $\mathcal{T}_{\lambda\mu}$. By the previous lemma, it suffices to show that $\bar{\theta}_{T_1} e_t, \bar{\theta}_{T_2} e_t, \ldots, \bar{\theta}_{T_m} e_t$ are independent, where t is our fixed tableau. For all i we have

$$\bar{\theta}_{T_i} e_t = \theta_{T_i} \kappa_t \{t\} = \kappa_t \theta_{T_i} \{t\}.$$

Now T_i is semistandard, so $[T_i] \rhd [S]$ for any other summand S in $\theta_{T_i} \{t\}$ (Corollary 2.10.3). The same is true for summands of $\kappa_t \theta_{T_i} \{t\}$, since the permutations in κ_t do not change the column equivalence class. Also, the $[T_i]$ are all distinct, since no equivalence class has more than one semistandard tableau. Hence the $\kappa_t \theta_{T_i} \{t\} = \bar{\theta}_{T_i} e_t$ satisfy the hypotheses of Lemma 2.10.4, making them independent. ∎

To prove that the $\bar{\theta}_T$ span, we need a lemma.

Lemma 2.10.7 *Consider $\theta \in \mathrm{Hom}(S^\lambda, M^\mu)$. Write*

$$\theta e_t = \sum_T c_T T,$$

where t is the fixed tableau of shape λ.

1. *If $\pi \in C_t$ and $T_1 = \pi T_2$, then $c_{T_1} = (\mathrm{sgn}\,\pi) c_{T_2}$.*

2. *Every T_1 with a repetition in some column has $c_{T_1} = 0$.*

3. *If θ is not the zero map, then there exists a semistandard T_2 having $c_{T_2} \neq 0$.*

Proof. 1. Since $\pi \in C_t$, we have

$$\pi(\theta e_t) = \theta(\pi \kappa_t \{t\}) = \theta((\operatorname{sgn} \pi) \kappa_t \{t\}) = (\operatorname{sgn} \pi)(\theta e_t).$$

Therefore, $\theta e_t = \sum_T c_T T$ implies

$$\pi \sum_T c_T T = \pi(\theta e_t) = (\operatorname{sgn} \pi)(\theta e_t) = (\operatorname{sgn} \pi) \sum_T c_T T.$$

Comparing coefficients of πT_2 on the left and T_1 on the right yields $c_{T_2} = (\operatorname{sgn} \pi) c_{T_1}$, which is equivalent to part 1.

2. By hypothesis, there exists $(i,j) \in C_t$ with $(i,j)T_1 = T_1$. But then $c_{T_1} = -c_{T_1}$ by what we just proved, forcing this coefficient to be zero.

3. Since $\theta \neq 0$, we can pick T_2 with $c_{T_2} \neq 0$ such that $[T_2]$ is maximal. We claim that T_2 can be taken to be semistandard. By parts 1 and 2, we can choose T_2 so that its columns strictly increase.

Suppose, toward a contradiction, that we have a descent in row i. Thus T_2 has a pair of columns that look like

$$
\begin{array}{ccc}
& b_1 & \\
& \wedge & \\
& b_2 & \\
& \wedge & \\
& \vdots & \\
& \wedge & \\
a_i & > & b_i \\
\wedge & & \\
\vdots & & \\
\wedge & & \\
a_p & &
\end{array}
$$

Choose A and B as usual, and let $g_{A,B} = \sum_\pi (\operatorname{sgn} \pi)\pi$ be the associated Garnir element. We have

$$g_{A,B}\left(\sum_T c_T T\right) = g_{A,B}(\theta e_t) = \theta(g_{A,B} e_t) = \theta(0) = 0.$$

Now T_2 appears in $g_{A,B} T_2$ with coefficient 1 (since the permutations in $g_{A,B}$ move distinct elements of T_2). So to cancel T_2 in the previous equation, there must be a $T \neq T_2$ with $\pi T = T_2$ for some π in $g_{A,B}$. Thus T is just T_2 with some of the a_j's and b_k's exchanged. But then $[T] \rhd [T_2]$ by the dominance lemma for generalized tableaux (Lemma 2.10.2). This contradicts our choice of T_2. ∎

We are now in a position to prove that the $\bar{\theta}_T$ generate $\operatorname{Hom}(S^\lambda, M^\mu)$.

Proposition 2.10.8 *The set*

$$\{\bar{\theta}_T \ : \ T \in \mathcal{T}^0_{\lambda\mu}\}$$

spans $\mathrm{Hom}(S^\lambda, M^\mu)$.

Proof. Pick any $\theta \in \mathrm{Hom}(S^\lambda, M^\mu)$ and write

$$\theta e_t = \sum_T c_T \boldsymbol{T}. \qquad (2.13)$$

Consider

$$L_\theta = \{S \in \mathcal{T}^0_{\lambda\mu} \ : \ [S] \trianglelefteq [T] \text{ for some } T \text{ appearing in } \theta e_t\}.$$

In poset terminology, L_θ corresponds to the *lower order ideal* generated by the T in θe_t. We prove this proposition by induction on $|L_\theta|$.

If $|L_\theta| = 0$, then θ is the zero map by part 3 of the previous lemma. Such a θ is surely generated by our set!

If $|L_\theta| > 0$, then in equation (2.13) we can find a semistandard T_2 with $c_{T_2} \neq 0$. Furthermore, it follows from the proof of part 3 in Lemma 2.10.7 that we can choose $[T_2]$ maximal among those tableaux that appear in the sum. Now consider

$$\theta_2 = \theta - c_{T_2} \bar{\theta}_{T_2}.$$

We claim that L_{θ_2} is a subset of L_θ with T_2 removed. First of all, every S appearing in $\bar{\theta}_{T_2} e_t$ satisfies $[S] \trianglelefteq [T_2]$ (see the comment after Corollary 2.10.3), so $L_{\theta_2} \subseteq L_\theta$. Furthermore, by part 1 of Lemma 2.10.7, every S with $[S] = [T_2]$ appears with the same coefficient in θe_t and $c_{T_2} \bar{\theta}_{T_2} e_t$. Thus $T_2 \notin L_{\theta_2}$, since $[T_2]$ is maximal. By induction, θ_2 is in the span of the $\bar{\theta}_T$ and thus θ is as well.

This completes the proof of the proposition and of Theorem 2.10.1. ∎

2.11 Kostka Numbers and Young's Rule

The Kostka numbers count semistandard tableaux.

Definition 2.11.1 The *Kostka numbers* are

$$K_{\lambda\mu} = |\mathcal{T}^0_{\lambda\mu}|. \quad \blacksquare$$

As an immediate corollary of the semistandard basis theorem (Theorem 2.10.1), we have Young's rule.

Theorem 2.11.2 (Young's Rule) *The multiplicity of S^λ in M^μ is equal to the number of semistandard tableaux of shape λ and content μ, i.e.,*

$$M^\mu \cong \bigoplus_\lambda K_{\lambda\mu} S^\lambda. \quad \blacksquare$$

Note that by Corollary 2.4.7, we can restrict this direct sum to $\lambda \trianglerighteq \mu$. Let us look at some examples.

Example 2.11.3 Suppose $\mu = (2,2,1)$. Then the possible $\lambda \trianglerighteq \mu$ and the associated λ-tableaux of type μ are as follows:

$$\lambda^1 = (2,2,1) \quad \lambda^2 = (3,1,1) \quad \lambda^3 = (3,2) \quad \lambda^4 = (4,1) \quad \lambda^5 = (5)$$

$$
\begin{array}{ccccc}
\bullet\ \bullet & \bullet\ \bullet\ \bullet & \bullet\ \bullet\ \bullet & \bullet\ \bullet\ \bullet\ \bullet & \bullet\ \bullet\ \bullet\ \bullet\ \bullet \\
= \bullet\ \bullet & = \bullet & = \bullet\ \bullet & = \bullet & = \\
\bullet & \bullet & & &
\end{array}
$$

$$
\begin{array}{ccccc}
T:\ 1\ 1 & 1\ 1\ 2 & 1\ 1\ 2 & 1\ 1\ 2\ 2 & 1\ 1\ 2\ 2\ 3 \\
2\ 2 & 2 & 2\ 3 & 3 & \\
3 & 3 & & &
\end{array}
$$

$$
\begin{array}{cc}
1\ 1\ 3 & 1\ 1\ 2\ 3 \\
2\ 2 & 2
\end{array}
$$

Thus

$$M^{(2,2,1)} \cong S^{(2,2,1)} \oplus S^{(3,3,1)} \oplus 2S^{(3,2)} \oplus 2S^{(4,1)} \oplus S^{(5)}.$$

Example 2.11.4 For any μ, $K_{\mu\mu} = 1$. This is because the only μ-tableau of content μ is the one with all the 1's in row 1, all the 2's in row 2, etc. Of course, we saw this result in Corollary 2.4.7.

Example 2.11.5 For any μ, $K_{(n)\mu} = 1$. Obviously there is only one way to arrange a collection of numbers in weakly increasing order. It is also easy to see from a representation-theoretic viewpoint that M^μ must contain exactly one copy of $S^{(n)}$ (see Exercise 5b).

Example 2.11.6 For any λ, $K_{\lambda(1^n)} = f^\lambda$ (the number of standard tableaux of shape λ). This says that

$$M^{(1^n)} \cong \bigoplus_\lambda f^\lambda S^\lambda.$$

But $M^{(1^n)}$ is just the regular representation (Example 2.1.7) and $f^\lambda = \dim S^\lambda$ (Theorem 2.5.2). Thus this is merely the special case of Proposition 1.10.1, part 1, where $G = S_n$.

2.12 Exercises

1. Let ϕ^λ be the character of M^λ. Find (with proof) a formula for ϕ^λ_λ, the value of ϕ^λ on the conjugacy class K_λ.

2. Verify the details in Theorem 2.1.12.

3. Let $\lambda = (\lambda_1, \lambda_2, \ldots, \lambda_l)$ and $\mu = (\mu_1, \mu_2, \ldots, \mu_m)$ be partitions. Characterize the fact that λ is covered by μ if the ordering used is

    (a) lexicographic,

    (b) dominance.

4. Consider $S^{(n-1,1)}$, where each tabloid is identified with the element in its second row. Prove the following facts about this module and its character.

    (a) We have

    $$S^{(n-1,1)} = \{c_1 \mathbf{1} + c_2 \mathbf{2} + \cdots + c_n \mathbf{n} \; : \; c_1 + c_2 + \cdots + c_n = 0\}.$$

    (b) For any $\pi \in S_n$,

    $$\chi^{(n-1,1)}(\pi) = (\text{number of fixedpoints of } \pi) - 1.$$

5. Let the group G act on the set S. We say that G is *transitive* if, given any $s, t \in S$, there is a $g \in G$ with $gs = t$. The group is *doubly transitive* if, given any $s, t, u, v \in S$ with $s \neq u$ and $t \neq v$, there is a $g \in G$ with $gs = t$ and $gu = v$. Show the following.

    (a) The orbits of G's action partition S.

    (b) The multiplicity of the trivial representation in $V = \mathbb{C}S$ is the number of orbits. Thus if G is transitive, then the trivial representation occurs exactly once. What does this say about the module M^μ?

    (c) If G is doubly transitive and V has character χ, then $\chi - 1$ is an irreducible character of G. *Hint:* Fix $s \in S$ and use Frobenius reciprocity on the stabilizer $G_s \leq G$.

    (d) Use part (c) to conclude that in S_n the function

    $$f(\pi) = (\text{number of fixedpoints of } \pi) - 1$$

    is an irreducible character.

6. Show that every irreducible character of S_n is an integer-valued function.

7. Define a lexicographic order on tabloids as follows. Associate with any $\{t\}$ the composition $\lambda = (\lambda_1, \lambda_2, \ldots, \lambda_n)$, where λ_i is the number of the row containing $n - i + 1$. If $\{s\}$ and $\{t\}$ have associated compositions λ and μ, respectively, then $\{s\} \leq \{t\}$ in *lexicographic order* (also called *last letter order*) if $\lambda \leq \mu$.

    (a) Show that $\{s\} \trianglelefteq \{t\}$ implies $\{s\} \leq \{t\}$.

    (b) Repeat Exercise 3 with tabloids in place of partitions.

8. Verify that the permutations π chosen after Definition 2.6.1 do indeed form a transversal for $\mathcal{S}_A \times \mathcal{S}_B$ in $\mathcal{S}_{A \cup B}$.

9. Verify the statements made in case 2 for the computation of Young's natural representation (page 74).

10. In $\mathcal{S}_n$ consider the transpositions $\tau_k = (k, k+1)$ for $k = 1, \ldots, n-1$.

    (a) Prove that the τ_k generate $\mathcal{S}_n$ subject to the Coxeter relations

    $$
    \begin{aligned}
    \tau_k^2 &= \epsilon, & 1 &\le k \le n-1, \\
    \tau_k \tau_{k+1} \tau_k &= \tau_{k+1} \tau_k \tau_{k+1}, & 1 &\le k \le n-2, \\
    \tau_k \tau_l &= \tau_l \tau_k, & 1 &\le k, l \le n-1 \text{ and } |k - l| \ge 2.
    \end{aligned}
    $$

    (b) Show that if G_n is a group generated by g_k for $k = 1, \ldots, n-1$ subject to the relations above (replacing τ_k by g_k), then $G_n \cong \mathcal{S}_n$. *Hint:* Induct on n using cosets of the subgroup generated by $g_1, \ldots, g_{n-2}$.

11. Fix a partition λ and fix an ordering of standard λ-tableaux $t_1, t_2, \ldots$. Define the *axial distance* from k to $k+1$ in tableau t_i to be

    $$
    \delta_i = \delta_i(k, k+1) = (c' - r') - (c - r),
    $$

    where c, c' and r, r' are the column and row coordinates of k and $k+1$, respectively, in t_i. *Young's seminormal form* assigns to each transposition $\tau = (k, k+1)$ the matrix $\rho_\lambda(\tau)$ with entries

    $$
    \rho_\lambda(\tau)_{i,j} = \begin{cases}
    1/\delta_i & \text{if } i = j, \\
    1 - 1/\delta_i^2 & \text{if } \tau t_i = t_j \text{ and } i < j, \\
    1 & \text{if } \tau t_i = t_j \text{ and } i > j, \\
    0 & \text{otherwise.}
    \end{cases}
    $$

    (a) Show that every row and column of $\rho_\lambda(\tau)$ has at most two nonzero entries.

    (b) Show that ρ_λ can be extended to a representation of $\mathcal{S}_n$, where $\lambda \vdash n$, by using the Coxeter relations of Exercise 10.

    (c) Show that this representation is equivalent to the one afforded by S^λ.

12. All matrices for this problem have rows and columns indexed by partitions of n in dual lexicographic order. Define

    $$
    A = (|\mathcal{S}_\lambda \cap K_\mu|) \text{ and } B = (|\mathcal{S}_\mu| \cdot K_{\lambda\mu}).
    $$

    (Be sure to distinguish between the conjugacy class K_μ and the Kostka number $K_{\lambda\mu}$.) Show that A, B are upper triangular matrices and that $C = B(A^t)^{-1}$ is the character table of $\mathcal{S}_n$ (with the columns listed in reverse order). *Hint:* Use Frobenius reciprocity.
    Use this method to calculate the character table for $\mathcal{S}_4$.

13. Prove that, up to sign, the determinant of the character table for $\mathcal{S}_n$ is

$$\prod_{\lambda \vdash n} \prod_{\lambda_i \in \lambda} \lambda_i.$$

14. Prove the following results in two ways: once using representations and once combinatorially.

   (a) If $K_{\lambda\mu} \neq 0$, then $\lambda \trianglerighteq \mu$.

   (b) Suppose μ and ν are compositions with the same parts (only rearranged). Then for any λ, $K_{\lambda\mu} = K_{\lambda\nu}$. *Hint:* For the combinatorial proof, consider the case where μ and ν differ by an adjacent transposition of parts.

15. Let G be a group and let $H \leq G$ have index two. Prove the following.

   (a) H is normal in G.

   (b) Every conjugacy class of G having nonempty intersection with H becomes a conjugacy class of H or splits into two conjugacy classes of H having equal size. Furthermore, the conjugacy class K of G does not split in H if and only if some $k \in K$ commutes with some $g \notin H$.

   (c) Let χ be an irreducible character of G. Then $\chi{\downarrow}_H$ is irreducible or is the sum of two inequivalent irreducibles. Furthermore, $\chi{\downarrow}_H$ is irreducible if and only if $\chi(g) \neq 0$ for some $g \notin H$.

16. Let A_n denote the alternating subgroup of $\mathcal{S}_n$ and consider $\pi \in \mathcal{S}_n$ having cycle type $\lambda = (\lambda_1, \lambda_2, \ldots, \lambda_l)$.

   (a) Show that $\pi \in A_n$ if and only if $n - l$ is even.

   (b) Prove that the conjugacy classes of $\mathcal{S}_n$ that split in A_n are those where all parts of λ are odd and distinct.

17. Use the previous two exercises and the character table of $\mathcal{S}_4$ to find the character table of A_4.

Chapter 3

Combinatorial Algorithms

Many results about representations of the symmetric group can be approached in a purely combinatorial manner. The crucial link between these two viewpoints is the fact (Theorem 2.6.5, part 2) that the number of standard Young tableaux of given shape is the degree of the corresponding representation.

We begin this chapter with famed Robinson-Schensted algorithm [Rob 38, Sch 61], which provides a bijective proof of the identity

$$\sum_{\lambda \vdash n} (f^\lambda)^2 = n!$$

from part 3 of Theorem 2.6.5. This procedure has many surprising properties that are surveyed in the sections that follow. This includes a discussion of Schützenberger's jeu de taquin [Scü 76], which is a crucial tool for many results. In the last two sections we give a pair of formulae for

$$f^\lambda = \dim S^\lambda.$$

One is in terms of products, whereas the other involves determinants.

3.1 The Robinson-Schensted Algorithm

If we disregard its genesis, the identity

$$\sum_{\lambda \vdash n} (f^\lambda)^2 = n!$$

can be regarded as a purely combinatorial statement. It says that the number of elements in $\mathcal{S}_n$ is equal to the number of pairs of standard tableaux of the same shape λ as λ varies over all partitions of n. Thus it should be possible to give a purely combinatorial—i.e., bijective—proof of this formula. The Robinson-Schensted correspondence does exactly that. It was originally discovered by Robinson [Rob 38] and then found independently in quite a different form by Schensted [Sch 61]. It is the latter's version of the algorithm that we present.

The bijection is denoted by

$$\pi \overset{\text{R-S}}{\longleftrightarrow} (P, Q),$$

where $\pi \in S_n$ and P, Q are standard λ-tableaux, $\lambda \vdash n$. We first describe the map that, given a permutation, produces a pair of tableaux.

"$\pi \overset{\text{R-S}}{\longrightarrow} (P, Q)$" Suppose that π is given in two-line notation as

$$\pi = \begin{matrix} 1 & 2 & \cdots & n \\ x_1 & x_2 & \cdots & x_n \end{matrix}.$$

We construct a sequence of tableaux pairs

$$(P_0, Q_0) = (\emptyset, \emptyset), \ (P_1, Q_1), \ (P_2, Q_2), \ \ldots, \ (P_n, Q_n) = (P, Q), \qquad (3.1)$$

where $x_1, x_2, \ldots, x_n$ are *inserted* into the P's and $1, 2, \ldots, n$ are *placed* in the Q's so that sh $P_k = $ sh Q_k for all k. The operations of insertion and placement will now be described.

Let P be a *partial tableau*, i.e., an array with distinct entries whose rows and columns increase. (So a partial tableau will be standard if its elements are precisely $\{1, 2, \ldots, n\}$.) Also, let x be an element not in P. To *row insert x into P*, we proceed as follows (where := means replacement).

RS1 Set $R :=$ the first row of P.

RS2 **While** x is less than some element of row R, **do**

    RSa Let y be the smallest element of R greater than x and replace y by x in R (denoted by $R \leftarrow x$).

    RSb Set $x := y$ and $R :=$ the next row down.

RS3 Now x is greater than every element of R, so place x at the end of row R and **stop**.

To illustrate, suppose $x = 3$ and

$$P = \begin{matrix} 1 \, 2 \, 5 \, 8 \\ 4 \, 7 \\ 6 \\ 9 \end{matrix}.$$

To follow the *path of the insertion* of x into P, we put elements that are displaced (or *bumped*) during the insertion in boldface type:

$$
\begin{matrix}
1\,2\,5\,8 \leftarrow \mathbf{3} & 1\,2\,\mathbf{3}\,8 & & 1\,2\,3\,8 & & 1\,2\,3\,8 \\
4\,7 & 4\,7 & \leftarrow \mathbf{5} & 4\,\mathbf{5} & & 4\,5 \\
6 & 6 & & 6 & \leftarrow \mathbf{7} & 6\,\mathbf{7} \\
9 & 9 & & 9 & & 9
\end{matrix}
$$

If the result of row inserting x into P yields the tableau P', then write

$$r_x(P) = P'.$$

Note that the insertion rules have been carefully written so that P' still has increasing rows and columns.

Placement of an element in a tableau is even easier than insertion. Suppose that Q is a partial tableau of shape μ and that (i, j) is an outer corner of μ. If k is greater than every element of Q, then to *place* k *in* Q *at cell* (i, j), merely set $Q_{i,j} := k$. The restriction on k guarantees that the new array is still a partial tableau. For example, if we take

$$Q = \begin{matrix} 1 & 2 & 5 \\ 4 & 7 & \\ 6 & & \\ 8 & & \end{matrix} \quad ,$$

then placing $k = 9$ in cell $(i, j) = (2, 3)$ yields

$$\begin{matrix} 1 & 2 & 5 \\ 4 & 7 & 9 \\ 6 & & \\ 8 & & \end{matrix} \quad .$$

At last we can describe how to build the sequence (3.1) from the permutation

$$\pi = \begin{matrix} 1 & 2 & \cdots & n \\ x_1 & x_2 & \cdots & x_n \end{matrix} .$$

Start with a pair (P_0, Q_0) of empty tableaux. Assuming that (P_{k-1}, Q_{k-1}) has been constructed, define (P_k, Q_k) by

$$
\begin{aligned}
P_k &= r_{x_k}(P_{k-1}), \\
Q_k &= \text{place } k \text{ into } Q_{k-1} \text{ at the cell } (i, j) \text{ where the} \\
&\quad \text{insertion terminates.}
\end{aligned}
$$

Note that the definition of Q_k ensures that $\operatorname{sh} P_k = \operatorname{sh} Q_k$ for all k. We call $P = P_n$ the *P-tableau*, or *insertion tableau*, of π and write $P = P(\pi)$. Similarly, $Q = Q_n$ is called the *Q-tableau*, or *recording tableau*, and denoted by $Q = Q(\pi)$.

Now we consider an example of the complete algorithm. Boldface numbers are used for the elements of the lower line of π and hence also for the elements of the P_k. Let

$$\pi = \begin{matrix} 1 & 2 & 3 & 4 & 5 & 6 & 7 \\ \mathbf{4} & \mathbf{2} & \mathbf{3} & \mathbf{6} & \mathbf{5} & \mathbf{1} & \mathbf{7} \end{matrix} . \tag{3.2}$$

Then the tableaux constructed by the algorithm are

$$P_k : \quad \emptyset, \quad 4, \quad 2, \quad \begin{matrix} 2\,3 \\ 4 \end{matrix}, \quad \begin{matrix} 2\,3\,6 \\ 4 \end{matrix}, \quad \begin{matrix} 2\,3\,5 \\ 4\,6 \end{matrix}, \quad \begin{matrix} 1\,3\,5 \\ 2\,6 \\ 4 \end{matrix}, \quad \begin{matrix} 1\,3\,5\,7 \\ 2\,6 \\ 4 \end{matrix} \quad = P,$$

$$Q_k : \quad \emptyset, \quad 1, \quad \begin{matrix} 1 \\ 2 \end{matrix}, \quad \begin{matrix} 1\,3 \\ 2 \end{matrix}, \quad \begin{matrix} 1\,3\,4 \\ 2 \end{matrix}, \quad \begin{matrix} 1\,3\,4 \\ 2\,5 \end{matrix}, \quad \begin{matrix} 1\,3\,4 \\ 2\,5 \\ 6 \end{matrix}, \quad \begin{matrix} 1\,3\,4\,7 \\ 2\,5 \\ 6 \end{matrix} \quad = Q.$$

So

$$\begin{matrix} 1 & 2 & 3 & 4 & 5 & 6 & 7 \\ 4 & 2 & 3 & 6 & 5 & 1 & 7 \end{matrix} \xrightarrow{\text{R-S}} \left(\begin{matrix} 1\,3\,5\,7 \\ 2\,6 \\ 4 \end{matrix} \quad , \quad \begin{matrix} 1\,3\,4\,7 \\ 2\,5 \\ 6 \end{matrix} \right).$$

The main theorem about this correspondence is as follows.

Theorem 3.1.1 ([Rob 38, Sch 61]) *The map*

$$\pi \xrightarrow{\text{R-S}} (P, Q)$$

is a bijection between elements of S_n and pairs of standard tableaux of the same shape $\lambda \vdash n$.

Proof. To show that we have a bijection, it suffices to create an inverse.

"$(P, Q) \xrightarrow{\text{S-R}} \pi$" We merely reverse the preceding algorithm step by step. We begin by defining $(P_n, Q_n) = (P, Q)$. Assuming that (P_k, Q_k) has been constructed, we will find x_k (the kth element of π) and (P_{k-1}, Q_{k-1}). To avoid double subscripting in what follows, we use $P_{i,j}$ to stand for the (i, j) entry of P_k.

Find the cell (i, j) containing k in Q_k. Since this is the largest element in Q_k, $P_{i,j}$ must have been the last element to be displaced in the construction of P_k. We can now use the following procedure to *delete* $P_{i,j}$ from P. For convenience, we assume the existence of an empty zeroth row above the first row of P_k.

SR1 Set $x := P_{i,j}$ and erase $P_{i,j}$.
   Set $R :=$ the $(i-1)$st row of P_k.

SR2 **While** R is not the zeroth row of P_k, **do**

   SRa Let y be the largest element of R smaller than x and replace y by x in R.

   SRb Set $x := y$ and $R :=$ the next row up.

SR3 Now x has been removed from the first row, so set $x_k := x$.

It is easy to see that P_{k-1} is P_k after the deletion process just described is complete and Q_{k-1} is Q_k with the k erased. Continuing in this way, we eventually recover all the elements of π in reverse order. ∎

The Robinson-Schensted algorithm has many surprising and beautiful properties. The rest of this chapter is devoted to discussing some of them.

3.2 Column Insertion

Obviously, we can define *column insertion* of x into P by replacing *row* by *column* everywhere in RS1–3. If column insertion of x into P yields P', we write

$$c_x(P) = P'.$$

It turns out that the column and row insertion operators commute. Before we can prove this, however, we need a lemma about the insertion path. The reader should be able to supply the details of the proof.

Lemma 3.2.1 *Let P be a partial tableau with $x \notin P$. Suppose that during the insertion $r_x(P) = P'$, the elements $x', x'', x''', \ldots$ are bumped from cells $(1, j'), (2, j''), (3, j'''), \ldots$, respectively. Then*

1. $x < x' < x'' < \cdots$,

2. $j' \geq j'' \geq j''' \geq \cdots$,

3. $P'_{i,j} \leq P_{i,j}$ for all i, j. ∎

Proposition 3.2.2 ([Sch 61]) *For any partial tableau P and distinct elements $x, y \notin P$,*

$$c_y r_x(P) = r_x c_y(P).$$

Proof. Let m be the maximum element in $\{x, y\} \cup P$. Then m cannot displace any entry during any of the insertions. The proof breaks into cases depending on where m is located.

Case 1: $y = m$. (The case where $x = m$ is similar.) Represent P schematically as

$$P = \qquad\qquad . \qquad\qquad\qquad (3.3)$$

Since y is maximal, c_y applied to either P or $r_x(P)$ merely inserts y at the end of the first column. Let $\overline{x}$ be the last element to be bumped during the insertion $r_x(P)$ and suppose it comes to rest in cell u. If u is in the first column, then

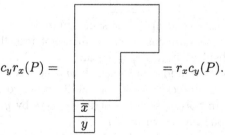

$$c_y r_x(P) = \qquad\qquad = r_x c_y(P).$$

If u is in any other column, then both $c_y r_x(P)$ and $r_x c_y(P)$ are equal to

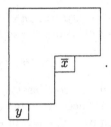

Case 2: $m \in P$. We induct on the number of entries in P. Let $\overline{P}$ be P with the m erased. Then $c_y r_x(\overline{P}) \subset c_y r_x(P)$ and $r_x c_y(\overline{P}) \subset r_x c_y(P)$. But $c_y r_x(\overline{P}) = r_x c_y(\overline{P})$ by induction. Thus $c_y r_x(P)$ and $r_x c_y(P)$ agree everywhere except possibly on the location of m.

To show that m occupies the same position in both tableaux, let $\overline{x}$ be the last element displaced during $r_x(\overline{P})$, say into cell u. Similarly, define $\overline{y}$ and v for the insertion $c_y(\overline{P})$. We now have two subcases, depending on whether u and v are equal or not.

Subcase 2a: $u = v$. Represent $\overline{P}$ as in diagram (3.3). Then $r_x(\overline{P})$ and $c_y(\overline{P})$ are represented, respectively, by

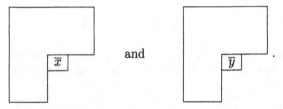

By the previous lemma, the first insertion displaces only elements in columns (weakly) to the right of $u = v$, and the second affects only those in rows (weakly) below. Thus r_x follows the same insertion path when applied to both $\overline{P}$ and $c_y(\overline{P})$ until $\overline{x}$ is bumped from the row above u. A similar statement holds for c_y in $\overline{P}$ and $r_x(\overline{P})$. Thus if $\overline{x} < \overline{y}$ (the case $\overline{x} > \overline{y}$ being similar), then we have

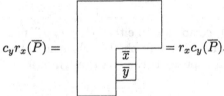

Note that $\overline{x}$ and $\overline{y}$ must end up in the same column.

Now consider what happens to m. If m is not in cell $u = v$, then none of the insertions displace it. So $c_y r_x(P)$ and $r_x c_y(P)$ are both equal to the previous diagram with m added in its cell. If m occupies u, then both $c_y r_x(P)$ and $r_x c_y(P)$ equal the preceding diagram with m added in the column just to the right of u. (In both cases m is bumped there by $\overline{y}$, and in the former it is first bumped one row down by $\overline{x}$.)

Subcase 2b: $u \neq v$. Let $\overline{P}$ have shape $\overline{\lambda}$ and let $c_y r_x(\overline{P}) = r_x c_y(\overline{P})$ have shape λ. Comparing $r_x(\overline{P})$ with $c_y r_x(\overline{P})$ and $c_y(\overline{P})$ with $r_x c_y(\overline{P})$, we see that $\overline{\lambda} \cup \{u\} \subset \lambda$ and $\overline{\lambda} \cup \{v\} \subset \lambda$. Thus

$$\lambda = \overline{\lambda} \cup \{u, v\}.$$

Since the insertion paths for $r_x(\overline{P})$ and $c_y(\overline{P})$ may cross, $\overline{x}$ and $\overline{y}$ are not necessarily the entries of cells u and v in $c_y r_x(\overline{P}) = r_x c_y(\overline{P})$. However, one can still verify that u is filled by r_x and v by r_y, whatever the order of the insertions. Consideration of all possible places where m could occur in P now completes the proof. The details are similar to the end of Subcase 2a and are left to the reader. ∎

As an application of this previous proposition, let us consider the *reversal* of π, denoted by π^r; i.e., if $\pi = x_1 x_2 \ldots x_n$, then $\pi^r = x_n x_{n-1} \ldots x_1$. The P-tableaux of π and π^r are closely related.

Theorem 3.2.3 ([Sch 61]) *If $P(\pi) = P$, then $P(\pi^r) = P^t$, where t denotes transposition.*

Proof. We have

$$
\begin{aligned}
P(\pi^r) &= r_{x_1} \cdots r_{x_{n-1}} r_{x_n}(\emptyset) && \text{(definition of } P(\pi^r)) \\
&= r_{x_1} \cdots r_{x_{n-1}} c_{x_n}(\emptyset) && \text{(initial tableau is empty)} \\
&= c_{x_n} r_{x_1} \cdots r_{x_{n-1}}(\emptyset) && \text{(Proposition 3.2.2)} \\
&\vdots \\
&= c_{x_n} c_{x_{n-1}} \cdots c_{x_1}(\emptyset) && \text{(induction)} \\
&= P^t. && \text{(definition of column insertion)} \blacksquare
\end{aligned}
$$

We will characterize the Q-tableau of π^r using Schützenberger's operation of evacuation (Theorem 3.9.4).

If we represent the elements of $\mathcal{S}_n$ as permutation matrices, then the dihedral group of the square (all reflections and rotations bringing a square back onto itself) acts on them. For example, reversal is just reflection in the vertical axis. It would be interesting to see what each of these changes in the permutation does to the output tableaux of the Robinson-Schensted algorithm. This theme motivates us in several of the following sections.

3.3 Increasing and Decreasing Subsequences

One of Schensted's main motivations for constructing his insertion algorithm, as the title of [Sch 61] suggests, was to study increasing and decreasing subsequences of a given sequence $\pi \in \mathcal{S}_n$. It turns out that these are intimately connected with the tableau $P(\pi)$.

Definition 3.3.1 Given $\pi = x_1 x_2 \ldots x_n \in \mathcal{S}_n$, an *increasing* (respectively, *decreasing*) *subsequence of π* is

$$x_{i_1} < x_{i_2} < \cdots < x_{i_k}$$

(respectively, $x_{i_1} > x_{i_2} > \cdots > x_{i_k}$) where $i_1 < i_2 < \cdots < i_k$. The integer k is the *length* of the subsequence. ∎

By way of illustration, take the permutation

$$\pi = 4\ 2\ 3\ 6\ 5\ 1\ 7. \tag{3.4}$$

Then an increasing subsequence of π of length 4 is 2 3 5 7, and a decreasing subsequence of length 3 is 4 3 1. In fact, it is easy to check that these are the longest increasing and decreasing subsequences of π. On page 94 we found the P-tableau of π to be

$$P(\pi) = \begin{array}{l} 1\ 3\ 5\ 7 \\ 2\ 6 \\ 4 \end{array} . \tag{3.5}$$

Note that the length of the first row of $P(\pi)$ is 4, whereas the length of the first column is 3. This is not a coincidence, as the next theorem shows.

Theorem 3.3.2 ([Sch 61]) *Consider* $\pi \in S_n$. *The length of the longest increasing subsequence of* π *is the length of the first row of* $P(\pi)$. *The length of the longest decreasing subsequence of* π *is the length of the first column of* $P(\pi)$.

Proof. Since reversing a permutation turns decreasing sequences into increasing ones, the second half of the theorem follows from the first and Theorem 3.2.3. To prove the first half, we actually demonstrate a stronger result. In what follows, P_{k-1} is the the tableau formed after $k-1$ insertions of the Robinson-Schensted algorithm.

Lemma 3.3.3 *If* $\pi = x_1 x_2 \ldots x_n$ *and* x_k *enters* P_{k-1} *in column* j, *then the longest increasing subsequence of* π *ending in* x_k *has length* j.

Proof. We induct on k. The result is trivial for $k = 1$, so suppose it holds for all values up to $k - 1$.

First we need to show the existence of an increasing subsequence of length j ending in x_k. Let y be the element of P_{k-1} in cell $(1, j-1)$. Then we have $y < x_k$, since x_k enters in column j. Also, by induction, there is an increasing subsequence σ of length $j-1$ ending in y. Thus σx_k is the desired subsequence.

Now we must prove that there cannot be a longer increasing subsequence ending in x_k. Suppose that such a subsequence exists and let x_i be the element preceding x_k in that subsequence. Then, by induction, when x_i is inserted, it enters in some column (weakly) to the right of column j. Thus the element y in cell $(1, j)$ of P_i satisfies

$$y \le x_i < x_k.$$

But by part 3 of Lemma 3.2.1, the entries in a given position of a tableau never increase with subsequent insertions. Thus the element in cell $(1, j)$ of

P_{k-1} is smaller than x_k, contradicting the fact that x_k displaces it. This finishes the proof of the lemma and hence of Theorem 3.3.2. ∎

Note that the first row of $P(\pi)$ need not be an increasing subsequence of π even though it has the right length; compare (3.4) and (3.5). However, an increasing subsequence of π of maximum length can be reconstructed from the Robinson-Schensted algorithm.

It turns out that an interpretation can be given to the lengths of the other rows and columns of $P(\pi)$, and we will do this in Section 3.5. First, however, we must develop an appropriate tool for the proof.

3.4 The Knuth Relations

Suppose we wish to prove a theorem about the elements of a set that is divided into equivalence classes. Then a common way to accomplish this is to show that

1. the theorem holds for a particular element of each equivalence class, and

2. if the theorem holds for one element of an equivalence class, then it holds for all elements of the class.

The set $\mathcal{S}_n$ has the following useful equivalence relation on it.

Definition 3.4.1 Two permutations $\pi, \sigma \in \mathcal{S}_n$ are said to be *P-equivalent*, written $\pi \stackrel{P}{\cong} \sigma$, if $P(\pi) = P(\sigma)$. ∎

For example, the equivalence classes in S_3 are

$$\{1\ 2\ 3\}, \quad \{2\ 1\ 3,\ 2\ 3\ 1\}, \quad \{1\ 3\ 2, 3\ 1\ 2\}, \quad \{3\ 2\ 1\},$$

corresponding to the P-tableaux

$$1\ 2\ 3, \qquad \begin{array}{l} 1\ 3 \\ 2 \end{array}, \qquad \begin{array}{l} 1\ 2 \\ 3 \end{array}, \qquad \begin{array}{l} 1 \\ 2 \\ 3 \end{array}.$$

respectively.

We can prove a strengthening of Schensted's Theorem (Theorem 3.3.2) using this equivalence relation and the proof technique just outlined. However, we also need an alternative description of P-equivalence. This is given by the Knuth relations.

Definition 3.4.2 Suppose $x < y < z$. Then $\pi, \sigma \in \mathcal{S}_n$ *differ by a Knuth relation of the first kind*, written $\pi \stackrel{1}{\cong} \sigma$, if

1. $\pi = x_1 \ldots yxz \ldots x_n$ and $\sigma = x_1 \ldots yzx \ldots x_n$ or vice versa.

They *differ by a Knuth relation of the second kind*, written $\pi \stackrel{2}{\cong} \sigma$, if

2. $\pi = x_1 \ldots xzy \ldots x_n$ and $\sigma = x_1 \ldots zxy \ldots x_n$ or vice versa.

The two permutations are *Knuth equivalent*, written $\pi \overset{K}{\cong} \sigma$, if there is a sequence of permutations such that

$$\pi = \pi_1 \overset{i}{\cong} \pi_2 \overset{j}{\cong} \cdots \overset{l}{\cong} \pi_k = \sigma,$$

where $i, j, \ldots, l \in \{1, 2\}$. ∎

Returning to S_3 we see that the only nontrivial Knuth relations are

$$2\,1\,3 \overset{1}{\cong} 2\,3\,1 \quad \text{and} \quad 1\,3\,2 \overset{2}{\cong} 3\,1\,2.$$

Thus the Knuth equivalence classes and P-equivalence classes coincide. This always happens.

Theorem 3.4.3 ([Knu 70]) *If $\pi, \sigma \in S_n$, then*

$$\pi \overset{K}{\cong} \sigma \iff \pi \overset{P}{\cong} \sigma.$$

Proof. "$\Longrightarrow$" It suffices to prove that $\pi \overset{P}{\cong} \sigma$ whenever π and σ differ by a Knuth relation. In fact, the result for $\overset{2}{\cong}$ follows from the one for $\overset{1}{\cong}$ because

$$
\begin{aligned}
\pi \overset{2}{\cong} \sigma \implies & \ \pi^r \overset{1}{\cong} \sigma^r && \text{(by definitions)} \\
\implies & \ P(\pi^r) = P(\sigma^r) && \text{(proved later)} \\
\implies & \ P(\pi)^t = P(\sigma)^t && \text{(Theorem 3.2.3)} \\
\implies & \ P(\pi) = P(\sigma).
\end{aligned}
$$

Now assume $\pi \overset{1}{\cong} \sigma$. Keeping the notation of Definition 3.4.2, let P be the tableau obtained by inserting the elements before y (which are the same in both π and σ). Thus it suffices to prove

$$r_z r_x r_y(P) = r_x r_z r_y(P).$$

Intuitively, what we will show is that the insertion path of y creates a "barrier," so that the paths for x and z lie to the left and right of this line, respectively, no matter in which order they are inserted. Since these two paths do not intersect, $r_z r_x$ and $r_x r_z$ have the same effect.

We induct on the number of rows of P. If P has no rows, i.e., $P = \emptyset$, then it is easy to verify that both sequences of insertions yield the tableau

$$
\begin{array}{cc}
x & z \\
y &
\end{array}
.
$$

Now let P have $r > 0$ rows and consider $\overline{P} = r_y(P)$. Suppose y enters P in column k, displacing element y'. The crucial facts about $\overline{P}$ are

1. $\overline{P}_{1,j} \leq y$ for all $j \leq k$, and

2. $\overline{P}_{1,l} > y'$ for all $l > k$.

If the insertion of x is performed next, then, since $x < y$, x must enter in some column j with $j \leq k$. Furthermore, if x' is the element displaced, then we must have $x' < y'$ because of our first crucial fact. Applying r_z to $r_x r_y(P)$ forces $z > y$ to come in at column l, where $l > k$ for the same reason. Also, the element z' displaced satisfies $z' > y'$ by crucial fact 2.

Now consider $r_x r_z r_y(P)$. Since the crucial facts continue to hold, z and x must enter in columns strictly to the right and weakly to the left of column k, respectively. Because these two sets of columns are disjoint, the entrance of one element does not affect the entrance of the other. Thus the first rows of $r_z r_x r_y(P)$ and $r_x r_z r_y(P)$ are equal. In addition, the elements displaced in both cases are x', y', and z', satisfying $x' < y' < z'$. Thus if P' denotes the bottom $r - 1$ rows of P, then the rest of our two tableaux can be found by computing $r_{z'} r_{x'} r_{y'}(P')$ and $r_{x'} r_{z'} r_{y'}(P')$. These last two arrays are equal by induction, so we are done with this half of the proof.

Before continuing with the other implication, we need to introduce some more concepts.

Definition 3.4.4 If P is a tableau, then the *row word of P* is the permutation

$$\pi_P = R_l R_{l-1} \ldots R_1,$$

where $R_1, \ldots, R_l$ are the rows of P. ∎

For example, if

$$P = \begin{array}{l} 1\ 3\ 5\ 7 \\ 2\ 6 \\ 4 \end{array} \quad ,$$

then

$$\pi_P = 4\ 2\ 6\ 1\ 3\ 5\ 7.$$

The following lemma is easy to verify directly from the definitions.

Lemma 3.4.5 *If P is a standard tableau, then*

$$\pi_P \xrightarrow{\text{R-S}} (P, \cdot). \quad ∎$$

The reader may have noticed that most of the definitions and theorems above involving standard tableaux and permutations make equally good sense when applied to partial tableaux and *partial permutations*, which are bijections $\pi : K \to L$ between two sets of positive integers. If $K = \{k_1 < k_2 < \cdots < k_m\}$, then we can write π in two-line form as

$$\pi = \begin{array}{cccc} k_1 & k_2 & \cdots & k_m \\ l_1 & l_2 & \cdots & l_m \end{array} ,$$

where $l_i = \pi(k_i)$. Insertion of the k_i and placement of the l_i set up a bijection $\pi \xrightarrow{\text{R-S}} (P, Q)$ with partial tableaux. Henceforth, we assume the more general case whenever it suits us to do so. Now back to the proof of Theorem 3.4.3.

"$\Longleftarrow$" By transitivity of equivalence relations, it suffices to show that

$$\pi \stackrel{K}{\cong} \pi_P$$

whenever $P(\pi) = P$.

We induct on n, the number of elements in π. Let x be the last element of π, so that $\pi = \pi'x$, where π' is a sequence of $n-1$ integers. Let $P' = P(\pi')$. Then, by induction,

$$\pi' \stackrel{K}{\cong} \pi_{P'}.$$

So it suffices to prove that

$$\pi_{P'}x \stackrel{K}{\cong} \pi_P.$$

In fact, we will show that the Knuth relations used to transform $\pi_{P'}x$ into π_P essentially simulate the insertion of x into the tableau P' (which of course yields the tableau P).

Let the rows of P' be $R_1, \ldots, R_l$, where $R_1 = p_1p_2 \ldots p_k$. If x enters P' in column j, then

$$p_1 < \cdots < p_{j-1} < x < p_j < \cdots < p_k.$$

Thus

$$
\begin{aligned}
\pi_{P'}x &= R_l \ldots R_2 p_1 \ldots p_{j-1}p_j \ldots p_{k-1}p_k x \\
&\stackrel{1}{\cong} R_l \ldots R_2 p_1 \ldots p_{j-1}p_j \ldots p_{k-1}x p_k \\
&\stackrel{1}{\cong} R_l \ldots R_2 p_1 \ldots p_{j-1}p_j \ldots x p_{k-1}p_k \\
&\quad \vdots \\
&\stackrel{1}{\cong} R_l \ldots R_2 p_1 \ldots p_{j-1}p_j x p_{j+1} \ldots p_k \\
&\stackrel{2}{\cong} R_l \ldots R_2 p_1 \ldots p_j p_{j-1}x p_{j+1} \ldots p_k \\
&\quad \vdots \\
&\stackrel{2}{\cong} R_l \ldots R_2 p_j p_1 \ldots p_{j-1}x p_{j+1} \ldots p_k.
\end{aligned}
$$

Now the tail of our permutation is exactly the first row of $P = r_x(P')$. Also, the element bumped from the first row of P', namely p_j, is poised at the end of the sequence corresponding to R_2. Thus we can model its insertion into the second row just as we did with x and R_1. Continuing in this manner, we eventually obtain the row word of P, completing the proof. ∎

3.5 Subsequences Again

Curtis Greene [Gre 74] gave a nice generalization of Schensted's result on increasing and decreasing sequences (Theorem 3.3.2). It involves unions of such sequences.

Definition 3.5.1 Let π be a sequence. A subsequence σ of π is *k-increasing* if, as a set, it can be written as

$$\sigma = \sigma_1 \uplus \sigma_2 \uplus \cdots \uplus \sigma_k,$$

where the σ_i are increasing subsequences of π. If the σ_i are all decreasing, then we say that σ is *k-decreasing*. Let

$$i_k(\pi) = \text{the length of } \pi\text{'s longest } k\text{-increasing subsequence}$$

and

$$d_k(\pi) = \text{the length of } \pi\text{'s longest } k\text{-decreasing subsequence.} \quad \blacksquare$$

Notice that the case $k = 1$ corresponds to sequences that are merely increasing or decreasing. Using our canonical example permutation (3.4),

$$\pi = 4\ 2\ 3\ 6\ 5\ 1\ 7,$$

we see that longest 1-, 2-, and 3-increasing subsequences are given by

$$2\ 3\ 5\ 7$$

and

$$4\ 2\ 3\ 6\ 5\ 7 \ =\ 2\ 3\ 5\ 7 \uplus 4\ 6$$

and

$$4\ 2\ 3\ 6\ 5\ 1\ 7 \ =\ 2\ 3\ 5\ 7 \uplus 4\ 6 \uplus 1,$$

respectively. Thus

$$i_1(\pi) = 4, \qquad i_2(\pi) = 6, \qquad i_3(\pi) = 7. \tag{3.6}$$

Recall that $P(\pi)$ in (3.5) had shape $\lambda = (4, 2, 1)$, so that

$$\lambda_1 = 4, \qquad \lambda_1 + \lambda_2 = 6, \qquad \lambda_1 + \lambda_2 + \lambda_3 = 7. \tag{3.7}$$

Comparing the values in (3.6) and (3.7), the reader should see a pattern.

To talk about the situation with k-decreasing subsequences conveniently, we need some notation concerning the columns of λ.

Definition 3.5.2 Let λ be a Ferrers diagram; then the *conjugate of λ* is

$$\lambda' = (\lambda_1', \lambda_2', \ldots, \lambda_m'),$$

where λ_i' is the length of the ith column of λ. Otherwise put, λ' is just the transpose of the diagram of λ and so it is sometimes denoted by λ^t. $\blacksquare$

As an example of the preceding definition, note that if

$$\lambda = (4,2,1) = \quad\begin{matrix} \bullet & \bullet & \bullet & \bullet \\ \bullet & \bullet & & \\ \bullet & & & \end{matrix} \quad,$$

then

$$\lambda' = \lambda^t = \quad\begin{matrix} \bullet & \bullet & \bullet \\ \bullet & \bullet & \\ \bullet & & \\ \bullet & & \end{matrix} \quad = (3,2,1,1).$$

Greene's theorem is as follows.

Theorem 3.5.3 ([Gre 74]) *Given $\pi \in \mathcal{S}_n$, let $\operatorname{sh} P(\pi) = (\lambda_1, \lambda_2, \ldots, \lambda_l)$ with conjugate $(\lambda_1', \lambda_2', \ldots, \lambda_m')$. Then for any k,*

$$\begin{aligned} i_k(\pi) &= \lambda_1 + \lambda_2 + \cdots + \lambda_k, \\ d_k(\pi) &= \lambda_1' + \lambda_2' + \cdots + \lambda_k'. \end{aligned}$$

Proof. By Theorem 3.2.3, it suffices to prove the statement for k-increasing subsequences.

We use the proof technique detailed at the beginning of Section 3.4 and the equivalence relation is the one studied there. Given an equivalence class corresponding to a tableau P, use the permutation π_P as the special representative of that class. So we must first prove the result for the row word.

By construction, the first k rows of P form a k-increasing subsequence of π_P, so $i_k(\pi_P) \geq \lambda_1 + \lambda_2 + \cdots + \lambda_k$. To show that the reverse inequality holds, note that any k-increasing subsequence can intersect a given decreasing subsequence in at most k elements. Since the columns of P partition π_P into decreasing subsequences,

$$i_k(\pi_P) \leq \sum_{i=1}^{m} \min(\lambda_i', k) = \sum_{i=1}^{k} \lambda_i.$$

Now we must show that the theorem holds for all permutations in the equivalence class of P. Given π in this class, Theorem 3.4.3 guarantees the existence of permutations $\pi_1, \pi_2, \ldots, \pi_j$ such that

$$\pi = \pi_1 \overset{K}{\cong} \pi_2 \overset{K}{\cong} \cdots \overset{K}{\cong} \pi_j = \pi_P,$$

where each π_{i+1} differs from π_i by a Knuth relation. Since all the π_i have the same P-tableau, we need to prove only that they all have the same value for i_k. Thus it suffices to show that if $\pi \overset{i}{\cong} \sigma$ for $i = 1, 2$, then $i_k(\pi) = i_k(\sigma)$. We will do the case of $\overset{1}{\cong}$, leaving $\overset{2}{\cong}$ as an exercise for the reader.

Suppose $\pi = x_1 \ldots yxz \ldots x_n$ and $\sigma = x_1 \ldots yzx \ldots x_n$. To prove that $i_k(\pi) \leq i_k(\sigma)$, we need to show only that any k-increasing subsequence of π has a corresponding k-increasing subsequence in σ of the same length. Let

$$\pi' = \pi_1 \uplus \pi_2 \uplus \cdots \uplus \pi_k$$

be the subsequence of π. If x and z are not in the same π_i for any i, then the π_i are also increasing subsequences of σ, and we are done.

Now suppose that x and z are both in π_1 (the choice of which of the component subsequences is immaterial). If $y \notin \pi'$, then let

$$\sigma_1 = \pi_1 \text{ with } x \text{ replaced by } y.$$

Since $x < y < z$, σ_1 is still increasing and

$$\sigma' = \sigma_1 \uplus \pi_2 \uplus \cdots \uplus \pi_k$$

is a subsequence of σ of the right length. Finally, suppose $y \in \pi'$, say $y \in \pi_2$ (note that we can not have $y \in \pi_1$). Let

$$
\begin{aligned}
\pi_1' &= \text{the subsequence of } \pi_1 \text{ up to and including } x, \\
\pi_1'' &= \text{the subsequence consisting of the rest of } \pi_1, \\
\pi_2' &= \text{the subsequence of } \pi_2 \text{ up to and including } y, \\
\pi_2'' &= \text{the subsequence consisting of the rest of } \pi_2.
\end{aligned}
$$

Note that $\pi_i = \pi_i' \pi_i''$ for $i = 1, 2$. Construct

$$\sigma_1 = \pi_1' \pi_2'' \quad \text{and} \quad \sigma_2 = \pi_2' \pi_1'',$$

which are increasing because $x < y < \min \pi_2''$ and $y < z$, respectively. Also,

$$\sigma' = \sigma_1 \uplus \sigma_2 \uplus \pi_3 \uplus \cdots \uplus \pi_k$$

is a subsequence of σ because x and z are no longer in the same component subsequence. Since its length is correct, we have proved the desired inequality.

To show that $i_k(\pi) \geq i_k(\sigma)$ is even easier. Since x and z are out of order in σ, they can never be in the same component of a k-increasing subsequence. Thus we are reduced to the first case, and this finishes the proof of the theorem. ∎

The fact that $i_k(\pi) = \lambda_1 + \lambda_2 + \cdots + \lambda_k$ does not imply that we can find a k-increasing subsequence of maximum length

$$\pi' = \pi_1 \uplus \pi_2 \uplus \cdots \uplus \pi_k$$

such that the length of π_i is λ_i for all i. As an example [Gre 74], consider

$$\pi = 2\ 4\ 7\ 9\ 5\ 1\ 3\ 6\ 8.$$

Then

$$
P(\pi) = \begin{matrix}
1 & 3 & 4 & 6 & 8 \\
2 & 4 & 9 & & \\
7 & & & &
\end{matrix} \quad,
$$

so $i_2(\pi) = 5 + 3 = 8$. There is only one 2-increasing subsequence of π having length 8, namely,

$$\pi' = 2\ 4\ 7\ 9 \uplus 1\ 3\ 6\ 8.$$

The reader can check that it is impossible to represent π' as the disjoint union of two increasing subsequences of lengths 5 and 3.

3.6 Viennot's Geometric Construction

We now return to our study (begun at the end of Section 3.2) of the effects that various changes in a permutation π have on the pair $(P(\pi), Q(\pi))$. We prove a remarkable theorem of Schützenberger [Scü 63] stating that taking the inverse of the permutation merely interchanges the two tableaux; i.e., if $\pi \overset{\text{R-S}}{\longrightarrow} (P, Q)$, then $\pi^{-1} \overset{\text{R-S}}{\longrightarrow} (Q, P)$. Our primary tool is a beautiful geometric description of the Robinson-Schensted correspondence due to Viennot [Vie 76].

Consider the first quadrant of the Cartesian plane. Given a permutation $\pi = x_1 x_2 \ldots x_n$, represent x_i by a box with coordinates (i, x_i) (compare this to the permutation matrix of π). Using our running example permutation from equation (3.4),

$$\pi = 4\ 2\ 3\ 6\ 5\ 1\ 7.$$

we obtain the following figure:

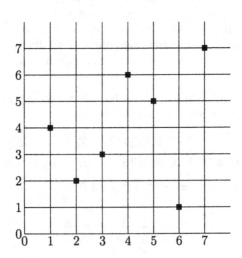

Now imagine a light shining from the origin so that each box casts a shadow with boundaries parallel to the coordinate axes. For example, the shadow cast by the box at $(4, 6)$ looks like this:

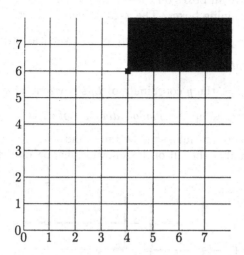

Similar figures result for the other boxes. Consider those points of the permutation that are in the shadow of no other point, in our case $(1, 4)$, $(2, 2)$, and $(6, 1)$. The *first shadow line, L_1*, is the boundary of the combined shadows of these boxes. In the next figure, the appropriate line has been thickened:

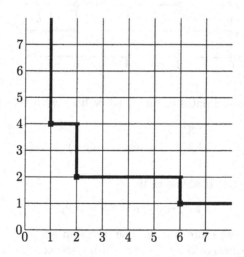

Note that this is a broken line consisting of line segments and exactly one horizontal and one vertical ray.

To form the *second shadow line, L_2*, one removes the boxes on the first shadow line and repeats this procedure.

Definition 3.6.1 Given a permutation displayed in the plane, we form its *shadow lines $L_1, L_2, \ldots$* as follows. Assuming that $L_1, \ldots, L_{i-1}$ have been constructed, remove all boxes on these lines. Then L_i is the boundary of the shadow of the remaining boxes. The *x-coordinate of L_i* is

$$x_{L_i} = \text{the } x\text{-coordinate of } L_i\text{'s vertical ray,}$$

and the *y-coordinate* is

$$y_{L_i} = \text{the } y\text{-coordinate of } L_i\text{'s horizontal ray.}$$

The shadow lines make up the *shadow diagram of π*. ∎

In our example there are four shadow lines, and their x- and y-coordinates are shown above and to the left of the following shadow diagram, respectively.

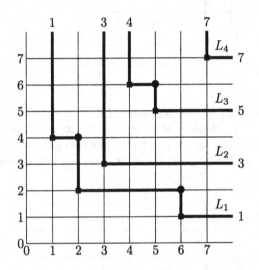

Compare the coordinates of our shadow lines with the first rows of the tableaux

$$P(\pi) = \begin{array}{l} 1\ 3\ 5\ 7 \\ 2\ 6 \\ 4 \end{array} \qquad \text{and} \qquad Q(\pi) = \begin{array}{l} 1\ 3\ 4\ 7 \\ 2\ 5 \\ 6 \end{array}$$

computed on page 94. It seems as if

$$P_{1,j} = y_{L_j} \text{ and } Q_{1,j} = x_{L_j} \qquad (3.8)$$

for all j. In fact, even more is true. The boxes on line L_j are precisely those elements passing through the $(1, j)$ cell during the construction of P, as the next result shows.

Lemma 3.6.2 *Let the shadow diagram of $\pi = x_1 x_2 \ldots x_n$ be constructed as before. Suppose the vertical line $x = k$ intersects i of the shadow lines. Let y_j be the y-coordinate of the lowest point of the intersection with L_j. Then the first row of the $P_k = P(x_1 \ldots x_k)$ is*

$$R_1 = y_1\ y_2\ \cdots\ y_i. \qquad (3.9)$$

Proof. Induct on k, the lemma being trivial for $k = 0$. Assume that the result holds for the line $x = k$ and consider $x = k + 1$. There are two cases.
    If

$$x_{k+1} > y_i, \tag{3.10}$$

then the box $(k + 1, x_{k+1})$ starts a new shadow line. So none of the values $y_1, \ldots, y_i$ change, and we obtain a new intersection,

$$y_{i+1} = x_{k+1}.$$

But by (3.9) and (3.10), the $(k + 1)$st intersection merely causes x_{k+1} to sit at the end of the first row without displacing any other element. Thus the lemma continues to be true.
    If, on the other hand,

$$y_1 < \cdots < y_{j-1} < x_{k+1} < y_j < \cdots < y_i, \tag{3.11}$$

then $(k + 1, x_{k+1})$ is added to line L_j. Thus the lowest coordinate on L_j becomes

$$y'_j = x_{k+1},$$

and all other y-values stay the same. Furthermore, equations (3.9) and (3.11) ensure that the first row of P_{k+1} is

$$y_1 \cdots y_{j-1} \, y'_j \cdots y_i.$$

This is precisely what is predicted by the shadow diagram. ∎

It follows from the proof of the previous lemma that the shadow diagram of π can be read left to right like a time-line recording the construction of $P(\pi)$. At the kth stage, the line $x = k$ intersects one shadow line in a ray or line segment and all the rest in single points. In terms of the first row of P_k, a ray corresponds to placing an element at the end, a line segment corresponds to displacing an element, and the points correspond to elements that are unchanged.

We can now prove that equation (3.8) always holds.

Corollary 3.6.3 ([Vie 76]) *If the permutation π has Robinson-Schensted tableaux (P, Q) and shadow lines L_j, then, for all j,*

$$P_{1,j} = y_{L_j} \quad and \quad Q_{1,j} = x_{L_j}.$$

Proof. The statement for P is just the case $k = n$ of Lemma 3.6.2.

As for Q, the entry k is added to Q in cell $(1, j)$ when x_k is greater than every element of the first row of P_{k-1}. But the previous lemma's proof shows that this happens precisely when the line $x = k$ intersects shadow line L_j in a vertical ray. In other words, $y_{L_j} = k = Q_{1,j}$ as desired. ∎

How do we recover the rest of the the P- and Q-tableaux from the shadow diagram of π? Consider the northeast corners of the shadow lines. These are marked with a dot in the diagram on page 108. If such a corner has coordinates (k, x'), then, by the proof of Lemma 3.6.2, x' must be displaced from the first row of P_{k-1} by the insertion of x_k. So the dots correspond to the elements inserted into the second row during the construction of P. Thus we can get the rest of the two tableaux by iterating the shadow diagram construction. In our example, the second and third rows come from the thickened and dashed lines, respectively, of the following diagram:

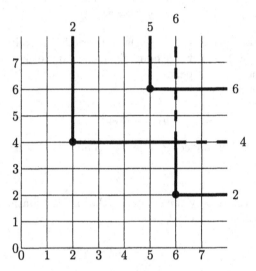

Formally, we have the following definition.

Definition 3.6.4 The *ith skeleton of* $\pi \in S_n$, $\pi^{(i)}$, is defined inductively by $\pi^{(1)} = \pi$ and

$$\pi^{(i)} = \begin{matrix} k_1 & k_2 & \cdots & k_m \\ l_1 & l_2 & \cdots & l_m \end{matrix},$$

where $(k_1, l_1), \ldots, (k_m, l_m)$ are the northeast corners of the shadow diagram of $\pi^{(i-1)}$ listed in lexicographic order. The shadow lines for $\pi^{(i)}$ are denoted by $L_j^{(i)}$. ∎

The next theorem should be clear, given Corollary 3.6.3 and the discussion surrounding Lemma 3.6.2.

Theorem 3.6.5 ([Vie 76]) *Suppose* $\pi \xrightarrow{\text{R-S}} (P, Q)$. *Then* $\pi^{(i)}$ *is a partial permutation such that*

$$\pi^{(i)} \xrightarrow{\text{R-S}} (P^{(i)}, Q^{(i)}),$$

where $P^{(i)}$ *(respectively,* $Q^{(i)}$*) consists of the rows* i *and below of* P *(respectively,* Q*). Furthermore,*

$$P_{i,j} = y_{L_j^{(i)}} \quad \text{and} \quad Q_{i,j} = x_{L_j^{(i)}}$$

for all i, j. ∎

It is now trivial to demonstrate Schützenberger's theorem.

Theorem 3.6.6 ([Scü 63]) *If $\pi \in S_n$, then*

$$P(\pi^{-1}) = Q(\pi) \quad and \quad Q(\pi^{-1}) = P(\pi).$$

Proof. Taking the inverse of a permutation corresponds to reflecting the shadow diagram in the line $y = x$. The theorem now follows from Theorem 3.6.5. ∎

As an application of Theorem 3.6.6 we can find those transpositions that, when applied to $\pi \in S_n$, leave $Q(\pi)$ invariant. Dual to our definition of P-equivalence is the following.

Definition 3.6.7 Two permutations $\pi, \sigma \in S_n$ are said to be *Q-equivalent*, written $\pi \overset{Q}{\cong} \sigma$, if $Q(\pi) = Q(\sigma)$. ∎

For example,

$$Q(2\ 1\ 3) = Q(3\ 1\ 2) = \begin{smallmatrix} 1 & 3 \\ 2 & \end{smallmatrix} \quad and \quad Q(1\ 3\ 2) = Q(2\ 3\ 1) = \begin{smallmatrix} 1 & 2 \\ 3 & \end{smallmatrix},$$

so

$$2\ 1\ 3 \overset{Q}{\cong} 3\ 1\ 2 \quad and \quad 1\ 3\ 2 \overset{Q}{\cong} 2\ 3\ 1. \tag{3.12}$$

We also have a dual notion for the Knuth relations.

Definition 3.6.8 Permutations $\pi, \sigma \in S_n$ *differ by a dual Knuth relation of the first kind*, written $\pi \overset{1^*}{\cong} \sigma$, if for some k,

1. $\pi = \ldots k+1 \ldots k \ldots k+2 \ldots$ and $\sigma = \ldots k+2 \ldots k \ldots k+1 \ldots$
    or vice versa.

They *differ by a dual Knuth relation of the second kind*, written $\pi \overset{2^*}{\cong} \sigma$, if for some k,

2. $\pi = \ldots k \ldots k+2 \ldots k+1 \ldots$ and $\sigma = \ldots k+1 \ldots k+2 \ldots k \ldots$
    or vice versa.

The two permutations are *dual Knuth equivalent*, written $\pi \overset{K^*}{\cong} \sigma$, if there is a sequence of permutations such that

$$\pi = \pi_1 \overset{i^*}{\cong} \pi_2 \overset{j^*}{\cong} \cdots \overset{l^*}{\cong} \pi_k = \sigma,$$

where $i, j, \ldots, l \in \{1, 2\}$. ∎

Note that the only two nontrivial dual Knuth relations in S_3 are

$$2\ 1\ 3 \stackrel{1^*}{\cong} 3\ 1\ 2 \quad \text{and} \quad 1\ 3\ 2 \stackrel{2^*}{\cong} 2\ 3\ 1.$$

These correspond exactly to (3.12).

The following lemma is obvious from the definitions. In fact, the definition of the dual Knuth relations was concocted precisely so that this result should hold.

Lemma 3.6.9 *If $\pi, \sigma \in S_n$, then*

$$\pi \stackrel{K}{\cong} \sigma \iff \pi^{-1} \stackrel{K^*}{\cong} \sigma^{-1}. \quad \blacksquare$$

Now it is an easy matter to derive the dual version of Knuth's theorem about P-equivalence (Theorem 3.4.3).

Theorem 3.6.10 *If $\pi, \sigma \in S_n$, then*

$$\pi \stackrel{K^*}{\cong} \sigma \iff \pi \stackrel{Q}{\cong} \sigma.$$

Proof. We have the following string of equivalences:

$$
\begin{aligned}
\pi \stackrel{K^*}{\cong} \sigma \quad &\iff \quad \pi^{-1} \stackrel{K}{\cong} \sigma^{-1} && \text{(Lemma 3.6.9)} \\
&\iff \quad P(\pi^{-1}) = P(\sigma^{-1}) && \text{(Theorem 3.4.3)} \\
&\iff \quad Q(\pi) = Q(\sigma). && \text{(Theorem 3.6.6)} \quad \blacksquare
\end{aligned}
$$

3.7   Schützenberger's Jeu de Taquin

The jeu de taquin (or "teasing game") of Schützenberger [Scü 76] is a powerful tool. It can be used to give alternative descriptions of both the P- and Q-tableaux of the Robinson-Schensted algorithm (Theorems 3.7.7 and 3.9.4) as well as the ordinary and dual Knuth relations (Theorems 3.7.8 and 3.8.8).

To get the full-strength version of these concepts, we must generalize to skew tableaux.

Definition 3.7.1 *If $\mu \subseteq \lambda$ as Ferrers diagrams, then the corresponding skew diagram, or skew shape, is the set of cells*

$$\lambda/\mu = \{c \,:\, c \in \lambda \text{ and } c \notin \mu\}.$$

A skew diagram is normal if $\mu = \emptyset$. $\blacksquare$

If $\lambda = (3,3,2,1)$ and $\mu = (2,1,1)$, then we have the skew diagram

$$\lambda/\mu = \qquad$$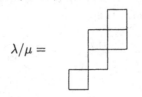

Of course, normal shapes are the left-justified ones we have been considering all along.

The definitions of skew tableaux, standard skew tableaux, and so on, are all as expected. In particular, the definition of the row word of a tableau still makes sense in this setting. Thus we can say that two skew partial tableaux P, Q are *Knuth equivalent*, written $P \overset{K}{\cong} Q$, if

$$\pi_P \overset{K}{\cong} \pi_Q.$$

Similar definitions hold for the other equivalence relations that we have introduced. Note that if $\pi = x_1 x_2 \ldots x_n$, then we can make π into a skew tableau by putting x_i in the cell $(n-i+1, i)$ for all i. This object is called the *antidiagonal strip tableau associated with* π and is also denoted by π. For example, if $\pi = 3142$ (a good approximation, albeit without the decimal point), then

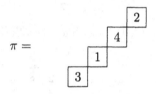

$$\pi =$$

So $\pi \overset{K}{\cong} \sigma$ as permutations if and only if $\pi \overset{K}{\cong} \sigma$ as tableaux.

We now come to the definition of a *jeu de taquin slide*, which is essential to all that follows.

Definition 3.7.2 Given a partial tableau P of shape λ/μ, we perform a *forward slide on P from cell c* as follows.

F1 Pick c to be an inner corner of μ.

F2 **While** c is not an inner corner of λ **do**

    Fa If $c = (i, j)$, then let c' be the cell of $\min\{P_{i+1,j}, P_{i,j+1}\}$.

    Fb Slide $P_{c'}$ into cell c and let $c := c'$.

If only one of $P_{i+1,j}, P_{i,j+1}$ exists in step Fa, then the maximum is taken to be that single value. We denote the resulting tableau by $j^c(P)$. Similarly, a *backward slide on P from cell c* produces a tableau $j_c(P)$ as follows

B1 Pick c to be an outer corner of λ.

B2 **While** c is not an outer corner of μ **do**

    Ba If $c = (i, j)$, then let c' be the cell of $\max\{P_{i-1,j}, P_{i,j-1}\}$.

    Bb Slide $P_{c'}$ into cell c and let $c := c'$. ∎

By way of illustration, let

$$P = \begin{array}{ccccc} & & 6 & 8 \\ & 2 & 4 & 5 & 9 \\ 1 & 3 & 7 & & \end{array}.$$

We let a dot indicate the position of the empty cell as we perform a forward slide from $c = (1,3)$.

$$\begin{array}{ccc}
\begin{array}{cccc} \bullet & 6 & 8 \\ 2 & 4 & 5 & 9 \\ 1 & 3 & 7 \end{array}, &
\begin{array}{cccc} 4 & 6 & 8 \\ 2 & \bullet & 5 & 9 \\ 1 & 3 & 7 \end{array}, &
\begin{array}{cccc} 4 & 6 & 8 \\ 2 & 5 & \bullet & 9 \\ 1 & 3 & 7 \end{array}, &
\begin{array}{cccc} 4 & 6 & 8 \\ 2 & 5 & 9 & \bullet \\ 1 & 3 & 7 \end{array}.
\end{array}$$

Thus

$$j^c(P) = \begin{array}{cccc} & 4 & 6 & 8 \\ & 2 & 5 & 9 \\ 1 & 3 & 7 \end{array}.$$

A backward slide from $c = (3,4)$ looks like the following.

$$\begin{array}{cccc}
\begin{array}{cccc} & 6 & 8 \\ 2 & 4 & 5 & 9 \\ 1 & 3 & 7 & \bullet \end{array}, &
\begin{array}{cccc} & 6 & 8 \\ 2 & 4 & 5 & 9 \\ 1 & 3 & \bullet & 7 \end{array}, &
\begin{array}{cccc} & 6 & 8 \\ 2 & \bullet & 5 & 9 \\ 1 & 3 & 4 & 7 \end{array}, &
\begin{array}{cccc} & 6 & 8 \\ \bullet & 2 & 5 & 9 \\ 1 & 3 & 4 & 7 \end{array}.
\end{array}$$

So

$$j_c(P) = \begin{array}{cccc} & 6 & 8 \\ & 2 & 5 & 9 \\ 1 & 3 & 4 & 7 \end{array}.$$

Note that a slide is an invertible operation. Specifically, if c is a cell for a forward slide on P and the cell vacated by the slide is d, then a backward slide into d restores P. In symbols,

$$j_d j^c(P) = P. \tag{3.13}$$

Similarly,

$$j^c j_d(P) = P. \tag{3.14}$$

if the roles of d and c are reversed.

Of course, we may want to make many slides in succession.

Definition 3.7.3 A sequence of cells $(c_1, c_2, \ldots, c_l)$ is a *slide sequence* for a tableau P if we can legally form $P = P_0, P_1, \ldots, P_l$, where P_i is obtained from P_{i-1} by performing a slide into cell c_i. Partial tableaux P and Q are *equivalent*, written $P \cong Q$, if Q can be obtained from P by some sequence of slides. ∎

This equivalence relation is the same as Knuth equivalence, as the next series of results shows.

Proposition 3.7.4 ([Scü 76]) *If P, Q are standard skew tableaux, then*

$$P \cong Q \Longrightarrow P \stackrel{K}{\cong} Q.$$

Proof. By induction, it suffices to prove the theorem when P and Q differ by a single slide. In fact, if we call the operation in steps Fb or Rb of the slide definition a *move*, then we need to demonstrate the result only when P and Q differ by a move. (The row word of a tableau with a hole in it can still be defined by merely ignoring the hole.)

The conclusion is trivial if the move is horizontal because then $\pi_P = \pi_Q$. If the move is vertical, then we can clearly restrict to the case where P and Q have only two rows. So suppose that x is the element being moved and that

$$P = \begin{array}{|c|c|c|c|}\hline R_l & x & & R_r \\\hline S_l & \bullet & S_r & \\\hline\end{array},$$

$$Q = \begin{array}{|c|c|c|c|}\hline R_l & \bullet & & R_r \\\hline S_l & x & S_r & \\\hline\end{array},$$

where R_l and S_l (respectively, R_r and S_r) are the left (respectively, right) portions of the two rows.

Now induct on the number of elements in P (or Q). If both tableaux consist only of x, then we are done.

Now suppose $|R_r| > |S_r|$. Let y be the rightmost element of R_r and let P', Q' be P, Q, respectively, with y removed. By our assumption P' and Q' are still skew tableaux, so applying induction yields

$$\pi_P = \pi_{P'} y \stackrel{K}{\cong} \pi_{Q'} y = \pi_Q.$$

The case $|S_l| > |R_l|$ is handled similarly.

Thus we are reduced to considering $|R_r| = |S_r|$ and $|R_l| = |S_l|$. Say

$$R_l = x_1 \ldots x_j, \qquad R_r = y_1 \ldots y_k,$$
$$S_l = z_1 \ldots z_j, \qquad S_r = w_1 \ldots w_k.$$

By induction, we may assume that one of j or k is positive. We will handle the situation where $j > 0$, leaving the other case to the reader. The following simple lemma will prove convenient.

Lemma 3.7.5 *Suppose $a_1 < a_2 < \cdots < a_n$.*

1. *If $x < a_1$, then $a_1 \ldots a_n x \stackrel{K}{\cong} a_1 x a_2 \ldots a_n$.*

2. *If $x > a_n$, then $x a_1 \ldots a_n \stackrel{K}{\cong} a_1 \ldots a_{n-1} x a_n$.* ∎

Since the rows and columns of P increase, we have $x_1 < z_i$ and $x_1 < w_i$ for all i as well as $x_1 < x$. Thus

$$
\begin{aligned}
\pi_P &= z_1 \ldots z_j w_1 \ldots w_k x_1 \ldots x_j x y_1 \ldots y_k \\
&\overset{K}{\cong} z_1 x_1 z_2 \ldots z_j w_1 \ldots w_k x_2 \ldots x_j x y_1 \ldots y_k \quad \text{(Lemma 3.7.5, part 1)} \\
&\overset{K}{\cong} z_1 x_1 z_2 \ldots z_j x w_1 \ldots w_k x_2 \ldots x_j y_1 \ldots y_k \quad \text{(induction)} \\
&\overset{K}{\cong} z_1 \ldots z_j x w_1 \ldots w_k x_1 \ldots x_j y_1 \ldots y_k \quad \text{(Lemma 3.7.5, part 1)} \\
&= \pi_Q. \qquad \blacksquare
\end{aligned}
$$

Schützenberger's teasing game can be described in the following manner.

Definition 3.7.6 Given a partial skew tableau P, we play *jeu de taquin* by choosing an arbitrary slide sequence that brings P to normal shape and then applying the slides. The resulting tableau is denoted by $j(P)$. ∎

It is not obvious at first blush that $j(P)$ is well defined—i.e., independent of the slide sequence. However, it turns out that we will always get the Robinson-Schensted P-tableau for the row word of P.

Theorem 3.7.7 ([Scü 76]) *If P is a partial skew tableau that is brought to a normal tableau P' by slides, then P' is unique. In fact, P' is the insertion tableau for π_P.*

Proof. By the previous proposition, $\pi_P \overset{K}{\cong} \pi_{P'}$. Thus by Knuth's theorem on P-equivalence (Theorem 3.4.3), π_P and $\pi_{P'}$ have the same insertion tableau. Finally, Lemma 3.4.5 tells us that the insertion tableau for $\pi_{P'}$ is just P' itself. ∎

We end this section by showing that equivalence and Knuth equivalence are indeed equivalent.

Theorem 3.7.8 ([Scü 76]) *Let P and Q be partial skew tableaux. Then*

$$
P \cong Q \iff P \overset{K}{\cong} Q.
$$

Proof. The only-if direction is Proposition 3.7.4. For the other implication, note that since $P \overset{K}{\cong} Q$, their row words must have the same P-tableau (Theorem 3.4.3 again). So by the previous theorem, $j(P) = j(Q) = P'$, say. Thus we can take P into Q by performing the slide sequence taking P to P' and then the inverse of the sequence taking Q to P'. Hence $P \cong Q$. ∎

3.8   Dual Equivalence

In the last section we gave a characterization of Knuth equivalence in terms of slides (Theorem 3.7.8). It would be nice to have a corresponding characterization of dual Knuth equivalence, and this was done by Haiman [Hai 92].

This machinery will also be useful when we prove the Littlewood-Richardson rule in Section 4.9. Before stating Haiman's key definition, we prove a useful result about normal tableaux.

Proposition 3.8.1 *Let P and Q be standard with the same normal shape* $\lambda \vdash n$. *Then $P \overset{K^*}{\cong} Q$.*

Proof. Induct on n, the proposition being trivial for $n \leq 2$. When $n \geq 3$, let c and d be the inner corner cells containing n in P and Q, respectively. There are two cases, depending on the relative positions of c and d.

If $c = d$, then let P' (respectively, Q') be P (respectively, Q) with the n erased. Now P' and Q' have the same shape, so by induction $\pi_{P'} \overset{K^*}{\cong} \pi_{Q'}$. But then we can apply the same sequence of dual Knuth relations to π_P and π_Q, the presence of n being immaterial. Thus $P \overset{K^*}{\cong} Q$ in this case.

If $c \neq d$, then it suffices to show the existence of two dual Knuth-equivalent tableaux P' and Q' with n in cells c and d, respectively. (Because then, by what we have shown in the first case, it follows that

$$\pi_P \overset{K^*}{\cong} \pi_{P'} \overset{K^*}{\cong} \pi_{Q'} \overset{K^*}{\cong} \pi_Q,$$

and we are done.) Let e be a lowest rightmost cell among all the cells on the boundary of λ between c and d. (The *boundary of λ* is the set of all cells at the end of a row or column of λ.) Schematically, we might have the situation

$$\lambda = \qquad \boxed{\begin{array}{c} \end{array}} \ \boxed{d}$$

Now let

$$\begin{array}{llll} P'_c & = & n, & P'_e & = & n-2, & P'_d & = & n-1; \\ Q'_c & = & n-1, & Q'_e & = & n-2, & Q'_d & = & n. \end{array}$$

and place the numbers $1, 2, \ldots, n-3$ anywhere as long as they are in the same cells in both P' and Q'. By construction, $\pi_P \overset{K^*}{\cong} \pi_Q$. ∎

The definition of dual equivalence is as follows.

Definition 3.8.2 Partial skew tableaux P and Q are *dual equivalent*, written $P \overset{*}{\cong} Q$, if whenever we apply the same slide sequence to both P and Q, we get resultant tableaux of the same shape. ∎

Note that the empty slide sequence can be applied to any tableau, so $P \overset{*}{\cong} Q$ implies that $\mathrm{sh}\, P = \mathrm{sh}\, Q$. The next result is proved directly from our definitions.

Lemma 3.8.3 *Let $P \stackrel{*}{\cong} Q$. If applying the same sequence of slides to both tableaux yields P' and Q', then $P' \stackrel{*}{\cong} Q'$.* ∎

One tableau may be used to determine a sequence of slides for another as follows. Let P and Q have shapes μ/ν and λ/μ, respectively. Then the cells of Q, taken in the order determined by Q's entries, are a sequence of backward slides on P. Let $j_Q(P)$ denote the result of applying this slide sequence to P. Also let $V = v_Q(P)$ stand for the tableau formed by the sequence of cells vacated during the construction of $j_Q(P)$, i.e.,

$$V_{i,j} = k \qquad \text{if } (i,j) \text{ was vacated when filling the cell of } k \in Q \qquad (3.15)$$

Displaying the elements of $\boldsymbol{P}$ as boldface and those of Q in normal type, we can compute an example.

$$
\begin{array}{c}
\begin{array}{ccc}
\mathbf{1} & \mathbf{2} & \mathbf{3} \\
\mathbf{4} & \mathbf{1} & \mathbf{3} \\
\mathbf{2} & &
\end{array} ,\
\begin{array}{ccc}
& \mathbf{2} & \mathbf{3} \\
\mathbf{1} & \mathbf{4} & \mathbf{3} \\
\mathbf{2} & &
\end{array} ,\
\begin{array}{ccc}
& \mathbf{2} & \mathbf{3} \\
& \mathbf{4} & \mathbf{3} \\
\mathbf{1} & &
\end{array} ,\
\begin{array}{ccc}
& & \mathbf{3} \\
\mathbf{2} & \mathbf{4} & \\
\mathbf{1} & &
\end{array}
= j_Q(\boldsymbol{P});
\end{array}
$$

$$
\begin{array}{c}
\varnothing \quad , \quad
\begin{array}{c}1\end{array} \quad , \quad
\begin{array}{c}1\\2\end{array} \quad , \quad
\begin{array}{cc}1 & 3\\2 &\end{array}
= v_Q(\boldsymbol{P}).
\end{array}
$$

With P and Q as before, the entries of P taken in reverse order define a sequence of forward slides on Q. Performing this sequence gives a tableau denoted by $j^P(Q)$. The vacating tableau $V = v^P(Q)$ is still defined by equation (3.15), with P in place of Q. The reader can check that using our example tableaux we obtain $j^P(Q) = v_Q(P)$ and $v^P(Q) = j_Q(P)$. This always happens, although it is not obvious. The interested reader should consult [Hai 92].

From these definitions it is clear that we can generalize equations (3.13) and (3.14). Letting $V = v^P(Q)$ and $W = v_Q(P)$, we have

$$j^W j_Q(P) = P \quad \text{and} \quad j_V j^P(Q) = Q. \qquad (3.16)$$

To show that dual equivalence and dual Knuth equivalence are really the same, we have to concentrate on small tableaux first.

Definition 3.8.4 *A partial tableau P is* miniature *if P has exactly three elements.* ∎

The miniature tableaux are used to model the dual Knuth relations of the first and second kinds.

Proposition 3.8.5 *Let P and Q be distinct miniature tableaux of the same shape λ/μ and content. Then*

$$P \stackrel{K^*}{\cong} Q \iff P \stackrel{*}{\cong} Q.$$

Proof. Without loss of generality, let P and Q be standard.

"$\Longrightarrow$" By induction on the number of slides, it suffices to show the following. Let c be a cell for a slide on P, Q and let P', Q' be the resultant tableaux. Then we must have

$$\operatorname{sh} P' = \operatorname{sh} Q' \text{ and } P' \stackrel{K^*}{\cong} Q'. \tag{3.17}$$

This is a tedious case-by-case verification. First, we must write down all the skew shapes with 3 cells (up to those translations that do not affect slides so the number of diagrams will be finite). Then we must find the possible tableau pairs for each shape (there will be at most two pairs corresponding to $\stackrel{1^*}{\cong}$ and $\stackrel{2^*}{\cong}$). Finally, all possible slides must be tried on each pair. We leave the details to the reader. However, we will do one of the cases as an illustration.

Suppose that $\lambda = (2, 1)$ and $\mu = \emptyset$. Then the only pair of tableaux of this shape is

$$P = \begin{array}{cc} 1 & 2 \\ 3 & \end{array} \quad \text{and} \quad Q = \begin{array}{cc} 1 & 3 \\ 2 & \end{array}$$

or vice versa, and $P \stackrel{1^*}{\cong} Q$. The results of the three possible slides on P, Q are given in the following table, from which it is easy to verify that (3.17) holds.

c :	$(1,3)$	$(2,2)$	$(3,1)$
P' :	$\begin{array}{cc} 1 & 2 \\ 3 & \end{array}$	$\begin{array}{cc} & 2 \\ 1 & 3 \end{array}$	$\begin{array}{c} 2 \\ 1 \\ 3 \end{array}$
Q' :	$\begin{array}{cc} 1 & 3 \\ 2 & \end{array}$	$\begin{array}{cc} & 1 \\ 2 & 3 \end{array}$	$\begin{array}{c} 3 \\ 1 \\ 2 \end{array}$
$\stackrel{i^*}{\cong}$:	$\stackrel{1^*}{\cong}$	$\stackrel{2^*}{\cong}$	$\stackrel{1^*}{\cong}$

"$\Longleftarrow$" Let N be a normal standard tableau of shape μ. So $P' = j^N(P)$ and $Q' = j^N(Q)$ are normal miniature tableaux. Now $P \stackrel{*}{\cong} Q$ implies that $\operatorname{sh} P' = \operatorname{sh} Q'$. This hypothesis also guarantees that $v^N(P) = v^N(Q) = V$, say. Applying equation (3.16),

$$j_V(P') = P \neq Q = j_V(Q'),$$

which gives $P' \neq Q'$. Thus P' and Q' are distinct miniature tableaux of the same normal shape. The only possibility is, then,

$$\{P', Q'\} = \left\{ \begin{array}{cc} 1 & 2 \\ 3 & \end{array}, \begin{array}{cc} 1 & 3 \\ 2 & \end{array} \right\}.$$

Since $P' \stackrel{K^*}{\cong} Q'$, we have, by what was proved in the forward direction,

$$P = j_V(P') \stackrel{K^*}{\cong} j_V(Q') = Q. \blacksquare$$

To make it more convenient to talk about miniature subtableaux of a larger tableau, we make the following definition.

Definition 3.8.6 Let P and Q be standard skew tableaux with

$$\mathrm{sh}\, P = \mu/\nu \vdash m \quad \text{and} \quad \mathrm{sh}\, Q = \lambda/\mu \vdash n.$$

Then $P \cup Q$ denotes the tableau of shape $\lambda/\nu \vdash m+n$ such that

$$(P \cup Q)_c = \begin{cases} P_c & \text{if } c \in \mu/\nu, \\ Q_c + m & \text{if } c \in \lambda/\mu. \end{cases} \blacksquare$$

Using the P and Q on page 118, we have

$$P \cup Q = \begin{array}{ccc} 1 & 2 & 3 \\ 4 & 5 & 7 \\ 6 & & \end{array}.$$

We need one more lemma before the main theorem of this section.

Lemma 3.8.7 ([Hai 92]) *Let V, W, P, and Q be standard skew tableaux with*

$$\mathrm{sh}\, V = \mu/\nu, \qquad \mathrm{sh}\, P = \mathrm{sh}\, Q = \lambda/\mu, \qquad \mathrm{sh}\, W = \kappa/\lambda.$$

Then

$$P \stackrel{*}{\cong} Q \implies V \cup P \cup W \stackrel{*}{\cong} V \cup Q \cup W.$$

Proof. Consider what happens in performing a single forward slide on $V \cup P \cup W$, say into cell c. Because of the relative order of the elements in the V, P, and W portions of the tableau, the slide can be broken up into three parts. First of all, the slide travels through V, creating a new tableau $V' = j_c(V)$ and vacating some inner corner d of μ. Then P becomes $P' = j_d(P)$, vacating cell e, and finally W is transformed into $W' = j_e(W)$. Thus $j_c(V \cup P \cup W) = V' \cup P' \cup W'$.

Now perform the same slide on $V \cup Q \cup W$. Tableau V is replaced by $j_c(V) = V'$, vacating d. If $Q' = j_d(Q)$, then, since $P \stackrel{*}{\cong} Q$, we have $\mathrm{sh}\, P' = \mathrm{sh}\, Q'$. So e is vacated as before, and W becomes W'. Thus $j_c(V \cup Q \cup W) = V' \cup Q' \cup W'$ with $P' \stackrel{*}{\cong} Q'$ by Lemma 3.8.3.

Now the preceding also holds, mutatis mutandis, to backward slides. Hence applying the same slide to both $V \cup P \cup W$ and $V \cup P \cup W$ yields tableaux of the same shape that still satisfy the hypotheses of the lemma. By induction, we are done. $\blacksquare$

We can now show that Proposition 3.8.5 actually holds for all pairs of tableaux.

Theorem 3.8.8 ([Hai 92]) *Let P and Q be standard tableaux of the same shape λ/μ. Then*

$$P \stackrel{K^*}{\cong} Q \iff P \stackrel{*}{\cong} Q.$$

Proof. "$\Longrightarrow$" We need to consider only the case where P and Q differ by a single dual Knuth relation, say the first (the second is similar). Now Q is obtained from P by switching $k+1$ and $k+2$ for some k. So

$$P = V \cup P' \cup W \text{ and } Q = V \cup Q' \cup W,$$

where P' and Q' are the miniature subtableaux of P and Q, respectively, that contain k, $k+1$, and $k+2$. By hypothesis, $P' \stackrel{1^*}{\cong} Q'$, which implies $P' \stackrel{*}{\cong} Q'$ by Proposition 3.8.5. But then the lemma just proved applies to show that $P \stackrel{*}{\cong} Q$.

"$\Longleftarrow$" Let tableau N be of normal shape μ. Let

$$P' = j^N(P) \text{ and } Q' = j^N(Q).$$

Since $P \stackrel{*}{\cong} Q$, we have $P' \stackrel{*}{\cong} Q'$ (Lemma 3.8.3) and $v^N(P) = v^N(Q) = V$ for some tableau V. Thus, in particular, $\operatorname{sh} P' = \operatorname{sh} Q'$, so that P' and Q' are dual Knuth equivalent by Proposition 3.8.1. Now, by definition, we have a sequence of dual Knuth relations

$$P' = P_1 \stackrel{i'^*}{\cong} P_2 \stackrel{j'^*}{\cong} \cdots \stackrel{l'^*}{\cong} P_k = Q',$$

where $i', j', \ldots, l' \in \{1, 2\}$. Hence the proof of the forward direction of Proposition 3.8.5 and equation (3.16) show that

$$P = j_V(P') \stackrel{i^*}{\cong} j_V(P_2) \stackrel{j^*}{\cong} \cdots \stackrel{l^*}{\cong} j_V(Q') = Q$$

for some $i, j, \ldots, l \in \{1, 2\}$. This finishes the proof of the theorem. ∎

3.9 Evacuation

We now return to our project of determining the effect that a reflection or rotation of the permutation matrix for π has on the tableaux $P(\pi)$ and $Q(\pi)$. We have already seen what happens when π is replaced by π^{-1} (Theorem 3.6.6). Also, Theorem 3.2.3 tells us what the P-tableau of π^r looks like. Since these two operations correspond to reflections that generate the dihedral group of the square, we will be done as soon as we determine $Q(\pi^r)$. To do this, another concept, called evacuation [Scü 63], is needed.

Definition 3.9.1 Let Q be a partial skew tableau and let m be the minimal element of Q. Then the *delta operator applied to Q* yields a new tableau, ΔQ, defined as follows.

D1 Erase m from its cell, c, in Q.

D2 Perform the slide j^c on the resultant tableau.

If Q is standard with n elements, then the *evacuation tableau for Q* is the vacating tableau $V = \mathrm{ev}\, Q$ for the sequence

$$Q,\ \Delta Q,\ \Delta^2 Q,\ \ldots,\ \Delta^n Q.$$

That is,

$$V_d = n - i \text{ if cell } d \text{ was vacated when passing from } \Delta^i Q \text{ to } \Delta^{i+1} Q. \ \blacksquare$$

Taking

$$Q = \begin{array}{l} 1\ 3\ 4\ 7 \\ 2\ 5 \\ 6 \end{array}$$

we compute $\mathrm{ev}\, Q$ as follows, using dots as placeholders for cells not yet filled.

$$\Delta^i Q: \begin{array}{llllllll} 1\ 3\ 4\ 7, & 2\ 3\ 4\ 7, & 3\ 4\ 7, & 4\ 7, & 5\ 7, & 6\ 7, & 7 \\ 2\ 5 & 5 & 5 & 5 & 6 & & \\ 6 & 6 & 6 & 6 & & & \end{array} \quad ;$$

$$\mathrm{ev}\, Q: \begin{array}{llllllll} \bullet\ \bullet\ \bullet\ \bullet, & \bullet\ \bullet\ \bullet\ \bullet, & \bullet\ \bullet\ \bullet\ 6, & \bullet\ \bullet\ 5\ 6, & \bullet\ \bullet\ 5\ 6, & \bullet\ \bullet\ 5\ 6, & \bullet\ 2\ 5\ 6 \\ \bullet\ \bullet & \bullet\ 7 & \bullet\ 7 & \bullet\ 7 & \bullet\ 7 & 3\ 7 & 3\ 7 \\ \bullet & \bullet & \bullet & \bullet & 4 & 4 & 4 \end{array} \ .$$

Completing the last slide we obtain

$$\mathrm{ev}\, Q = \begin{array}{llll} 1 & 2 & 5 & 6 \\ 3 & 7 & & \\ 4 & & & \end{array} \ .$$

Note that Q was the Q-tableau of our example permutation $\pi = 4\ 2\ 3\ 6\ 5\ 1\ 7$ in (3.4). The reader who wishes to anticipate Theorem 3.9.4 should now compute $Q(\pi^r)$ and compare the results.

Since the delta operator is defined in terms of slides, it is not surprising that it commutes with jeu de taquin.

Lemma 3.9.2 *Let Q be any skew partial tableau; then*

$$j\Delta(Q) = \Delta j(Q).$$

Proof. Let P be Q with its minimum element m erased from cell c. We write this as $P = Q - \{m\}$. Then

$$j\Delta(Q) = j(P)$$

by the definition of Δ and the uniqueness of the j operator (Theorem 3.7.7).

We now show that any forward slide on Q can be mimicked by a slide on P. Let $Q' = j^d(Q)$ for some cell d. There are two cases. If d is not vertically or horizontally adjacent to c, then it is legal to form $P' = j^d(P)$. Since m is involved in neither slide, we again have that $P' = Q' - \{m\}$.

If d is adjacent to c, then the first move of $j^d(Q)$ will put m in cell d and then proceed as a slide into c. Thus letting $P' = j^c(P)$ will preserve the relationship between P' and Q' as before.

Continuing in this manner, when we obtain $j(Q)$, we will have a corresponding P'', which is just $j(Q)$ with m erased from the $(1,1)$ cell. Thus

$$\Delta j(Q) = j^{(1,1)}(P'') = j(P) = j\Delta(Q). \blacksquare$$

As a corollary, we obtain our first relationship between the jeu de taquin and the Q-tableau of the Robinson-Schensted algorithm.

Proposition 3.9.3 ([Scü 63]) *Suppose* $\pi = x_1 x_2 \ldots x_n \in S_n$ *and let*

$$\bar{\pi} = \begin{matrix} 2 & 3 & \cdots & n \\ x_2 & x_3 & \cdots & x_n \end{matrix}.$$

Then

$$Q(\bar{\pi}) = \Delta Q(\pi).$$

Proof. Consider $\sigma = \pi^{-1}$ and $\bar{\sigma} = \bar{\pi}^{-1}$. Note that the lower line of $\bar{\sigma}$ is obtained from the lower line of σ by deleting the minimum element, 1.

By Theorem 3.6.6, it suffices to show that

$$P(\bar{\sigma}) = \Delta P(\sigma).$$

If we view σ and $\bar{\sigma}$ as antidiagonal strip tableaux, then $\bar{\sigma} \cong \Delta\sigma$. Hence by Theorem 3.7.7 and Lemma 3.9.2,

$$P(\bar{\sigma}) = P(\Delta\sigma) = \Delta P(\sigma). \blacksquare$$

Finally, we can complete our characterization of the image of π^r under the Robinson-Schensted map.

Theorem 3.9.4 ([Scü 63]) *If* $\pi \in S_n$, *then*

$$Q(\pi^r) = \text{ev } Q(\pi)^t.$$

Proof. Let $\pi, \bar{\pi}$ be as in the previous proposition with

$$\bar{\pi}^r = \begin{matrix} 1 & 2 & \cdots & n-1 \\ x_n & x_{n-1} & \cdots & x_2 \end{matrix}.$$

Induct on n. Now

$$
\begin{aligned}
Q(\pi^r) - \{n\} &= Q(\bar{\pi}^r) & (\bar{\pi}^r = x_n \ldots x_2) \\
&= \text{ev } Q(\bar{\pi})^t & \text{(induction)} \\
&= \text{ev } Q(\pi)^t - \{n\}. & \text{(Proposition 3.9.3)}
\end{aligned}
$$

Thus we need to show only that n occupies the same cell in both $Q(\pi^r)$ and $\text{ev}\, Q(\pi)^t$. Let $\text{sh}\, Q(\pi) = \lambda$ and $\text{sh}\, Q(\overline{\pi}) = \overline{\lambda}$. By Theorems 3.1.1 and 3.2.3 we have $\text{sh}\, Q(\pi^r) = \lambda^t$ and $\text{sh}\, Q(\overline{\pi}^r) = \overline{\lambda}^t$. Hence

$$
\begin{aligned}
\text{cell of } n \text{ in } Q(\pi^r) &= \lambda^t/\overline{\lambda}^t && (\overline{\pi}^r = x_n \ldots x_2) \\
&= (\lambda/\overline{\lambda})^t \\
&= (\text{cell of } n \text{ in ev}\, Q(\pi))^t. \quad (\text{Proposition 3.9.3}) \ \blacksquare
\end{aligned}
$$

3.10 The Hook Formula

There is an amazingly simple product formula for f^λ, the number of standard λ-tableaux. It involves objects called hooks.

Definition 3.10.1 If $v = (i, j)$ is a node in the diagram of λ, then it has *hook*

$$
H_v = H_{i,j} = \{(i, j') \; : \; j' \ge j\} \cup \{(i', j) \; : \; i' \ge i\}
$$

with corresponding *hooklength*

$$
h_v = h_{i,j} = |H_{i,j}|. \ \blacksquare
$$

To illustrate, if $\lambda = (4^2, 3^3, 1)$, then the dotted cells in

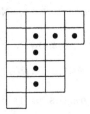

are the hook $H_{2,2}$ with hooklength $h_{2,2} = 6$.

It is now easy to state the hook formula of Frame, Robinson, and Thrall.

Theorem 3.10.2 (Hook Formula [FRT 54]) *If $\lambda \vdash n$, then*

$$
f^\lambda = \frac{n!}{\prod_{(i,j)\in\lambda} h_{i,j}}.
$$

Before proving this theorem, let us pause for an example and an anecdote. Suppose we wish to calculate the number of standard Young tableaux of shape $\lambda = (2, 2, 1) \vdash 5$. The hooklengths are given in the array

4	2
3	1
1	

where $h_{i,j}$ is placed in cell (i, j). Thus

$$
f^{(2,2,1)} = \frac{5!}{4 \cdot 3 \cdot 2 \cdot 1^2} = 5.
$$

This result can be verified by listing all possible tableaux:

$$\begin{array}{ccccc}
1\,2 & 1\,2 & 1\,3 & 1\,3 & 1\,4 \\
3\,4, & 3\,5, & 2\,4, & 2\,5, & 2\,5. \\
5 & 4 & 5 & 4 & 3
\end{array}$$

The tale of how the hook formula was born is an amusing one. One Thursday in May of 1953, Robinson was visiting Frame at Michigan State University. Discussing the work of Staal [Sta 50] (a student of Robinson), Frame was led to conjecture the hook formula. At first Robinson could not believe that such a simple formula existed, but after trying some examples he became convinced, and together they proved the identity. On Saturday they went to the University of Michigan, where Frame presented their new result after a lecture by Robinson. This surprised Thrall, who was in the audience, because he had just proved the same result on the same day!

There are many different bijective proofs of the hook formula. Franzblau and Zeilberger [F-Z 82] were the first to come up with a (complicated) bijection. Later, Zeilberger [Zei 84] gave a bijective version of a probabilistic proof of Greene, Nijenhuis, and Wilf [GNW 79] (see Exercise 17). But the map was still fairly complex. Also, Remmel [Rem 82] used the Garsia-Milne involution principle [G-M 81] to produce a bijection as a composition of maps. It was not until the 1990s that a truly straightforward one-to-one correspondence was found (even though the proof that it is correct is still somewhat involved), and that is the one we will present here. The algorithm was originally outlined by Pak and Stoyanovskii [PS 92], and then a complete proof was given by these two authors together with Novelli [NPS 97]. (A generalization of this method can be found in the work of Krattenthaler [Kra pr].)

To get the hook formula in a form suitable for bijective proof, we rewrite it as

$$n! = f^\lambda \prod_{(i,j)\in\lambda} h_{i,j}.$$

So it suffices to find a bijection

$$T \overset{\text{N-P-S}}{\longleftrightarrow} (P, J),$$

where $\text{sh}\,T = \text{sh}\,P = \text{sh}\,J = \lambda$, T is an arbitrary Young tableau, P is a standard tableau, and J is an array such that the number of choices for the entry $J_{i,j}$ is $h_{i,j}$. More specifically, define the *arm* and *leg* of $H_{i,j}$ to be

$$A_{i,j} = \{(i,j') \;:\; j' > j\} \quad \text{and} \quad L_{i,j} = \{(i',j) \;:\; i' > i\},$$

respectively, with corresponding *arm length* and *leg length*

$$al_{i,j} = |A_{i,j}| \quad \text{and} \quad ll_{i,j} = |L_{i,j}|.$$

In the previous example with $\lambda = (4^2, 3^3, 1)$ we have

$$\begin{array}{ll}
A_{2,2} = \{(2,3), (2,4)\}, & al_{2,2} = 2; \\
L_{2,2} = \{(3,2), (4,2), (5,2)\}; & ll_{2,2} = 3.
\end{array}$$

Note that $h_{i,j} = al_{i,j} + ll_{i,j} + 1$. So our requirement on the array J will be

$$-ll_{i,j} \leq J_{i,j} \leq al_{i,j}, \qquad (3.18)$$

and we will call J a *hook tableau*.

"$T \xrightarrow{\text{N-P-S}} (P, J)$" The basic idea behind this map is simple. We will use a modified jeu de taquin to unscramble the numbers in T so that rows and columns increase to form P. The hook tableau, J, will keep track of the unscrambling process so that T can be reconstructed from P.

We first need to totally order the cells of λ by defining

$$(i, j) \leq (i', j') \quad \text{if and only if} \quad j > j' \text{ or } j = j' \text{ and } i \geq i'.$$

Label the cells of λ in the given order

$$c_1 < c_2 < \ldots < c_n.$$

So, for example, if $\lambda = (3, 3, 2)$, then the ordering is

c_8	c_5	c_2
c_7	c_4	c_1
c_6	c_3	

.

Now if T is a λ-tableau and c is a cell, we let $T^{\leq c}$ (respectively, $T^{<c}$) be the tableau containing all cells b of T with $b \leq c$ (respectively, $b < c$). Continuing our example, if

$$T = \begin{matrix} 3 & 2 & 7 \\ 6 & 1 & 8 \\ 4 & 5 \end{matrix}, \quad \text{then} \quad T_{\leq c_6} = \begin{matrix} & 2 & 7 \\ & 1 & 8 \\ 4 & 5 \end{matrix}.$$

Given T, we will construct a sequence of pairs

$$(T_1, J_1) = (T, 0), \ (T_2, J_2), \ (T_3, J_3), \ \ldots, (T_n, J_n) = (P, J), \qquad (3.19)$$

where 0 is the tableau of all zeros and for all k we have that $T_k^{\leq c_k}$ is standard, and that J_k is a hook tableau which is zero outside $J_k^{\leq c_k}$.

If T is a tableau and c is a cell of T, then we perform a *modified forward slide* to form a new tableau $j^c(T)$ as follows.

NPS1 Pick c such that $T^{<c}$ is standard.

NPS2 **While** $T^{\leq c}$ is not standard **do**

NPSa If $c = (i, j)$, then let c' be the cell of $\min\{T_{i+1,j}, T_{i,j+1}\}$.

NPSb Exchange T_c and $T_{c'}$ and let $c := c'$.

It is easy to see that the algorithm will terminate since $T^{\leq c}$ will eventually become standard, in the worst case when c reaches an inner corner of λ. We call the sequence, p, of cells that c passes through the *path of c*. To illustrate, if

$$T = \begin{array}{cccc} 9 & 6 & 1 & 5 \\ 8 & 2 & 3 & 7 \\ 4 & & & \end{array}$$

and $c = (1,2)$, then here is the computation of $j^c(T)$, where the moving element is in boldface:

$$T = \begin{array}{cccc} 9 & \mathbf{6} & 1 & 5 \\ 8 & 2 & 3 & 7 \\ 4 & & & \end{array}, \quad \begin{array}{cccc} 9 & 1 & \mathbf{6} & 5 \\ 8 & 2 & 3 & 7 \\ 4 & & & \end{array}, \quad \begin{array}{cccc} 9 & 1 & 3 & 5 \\ 8 & 2 & \mathbf{6} & 7 \\ 4 & & & \end{array} = j^c(T).$$

So in this case the path of c is

$$p = ((1,2),\ (1,3),\ (2,3)).$$

We can now construct the sequence (3.19). The initial pair (T_1, J_1) is already defined. Assuming that we have (T_{k-1}, J_{k-1}) satisfying the conditions after (3.19), then let

$$T_k = j^{c_k}(T_{k-1}).$$

Furthermore, if j^{c_k} starts at $c_k = (i,j)$ and ends at (i', j'), then $J_k = J_{k-1}$ except for the values

$$(J_k)_{h,j} = \begin{cases} (J_{k-1})_{h+1,j} - 1 & \text{for } i \leq h < i', \\ j' - j & \text{for } h = i'. \end{cases}$$

It is not hard to see that if J_{k-1} was a hook tableau, then J_k will still be one. Starting with the tableau

$$T = \begin{array}{cc} 6 & 2 \\ 4 & 3 \\ 5 & 1 \end{array},$$

here is the whole algorithm.

$$T_k: \begin{array}{cc} 6 & 2 \\ 4 & 3 \\ 5 & 1 \end{array}, \begin{array}{cc} 6 & 2 \\ 4 & 1 \\ 5 & 3 \end{array}, \begin{array}{cc} 6 & 1 \\ 4 & 2 \\ 5 & 3 \end{array}, \begin{array}{cc} 6 & 1 \\ 4 & 2 \\ 3 & 5 \end{array}, \begin{array}{cc} 6 & 1 \\ 2 & 4 \\ 3 & 5 \end{array}, \begin{array}{cc} 1 & 4 \\ 2 & 5 \\ 3 & 6 \end{array} = P,$$

$$J_k: \begin{array}{cc} 0 & 0 \\ 0 & 0 \\ 0 & 0 \end{array}, \begin{array}{cc} 0 & 0 \\ 0 & -1 \\ 0 & 0 \end{array}, \begin{array}{cc} 0 & -2 \\ 0 & 0 \\ 0 & 0 \end{array}, \begin{array}{cc} 0 & -2 \\ 0 & 0 \\ 1 & 0 \end{array}, \begin{array}{cc} 0 & -2 \\ 1 & 0 \\ 1 & 0 \end{array}, \begin{array}{cc} 0 & -2 \\ 0 & 0 \\ 1 & 0 \end{array} = J.$$

Theorem 3.10.3 ([NPS 97]) *For fixed λ, the map*

$$T \xrightarrow{\text{N-P-S}} (P, J)$$

just defined is a bijection between tableaux T and pairs (P, J) with P a standard tableau and J a hook tableau such that $\operatorname{sh} T = \operatorname{sh} P = \operatorname{sh} J = \lambda$.

Proof. As usual, we will create an inverse.

"$(P,J) \xrightarrow{\text{S–P–N}} T$" To determine the sequence $(T_n, J_n), \ldots, (T_1, J_1)$ we start with $(T_n, J_n) = (P, J)$. Assuming that (T_k, J_k) has been constructed, we first need to consider which cells could have been the end of the path for $j^{c_k}(T_{k-1})$. The set of *candidate cells* for $c_k = (i_0, j_0)$ in T_k is

$$C_k = \{(i', j') \ : \ i' \geq i_0, \ j' = j_0 + (J_k)_{i', j_0}, \ (J_k)_{i', j_0} \geq 0\}.$$

So if we have

$$(T_{10}, J_{10}) = \begin{pmatrix} 13 & 18 & 10 & 1 & 4 & & 0 & 0 & 0 & -2 & -2 \\ 17 & 15 & 2 & 3 & 5 & & 0 & 0 & -1 & -2 & 0 \\ 14 & 16 & 6 & 7 & 11 & & 0 & 0 & 1 & 1 & 0 \\ 19 & 12 & 8 & 9 & & & 0 & 0 & 0 & 0 & \end{pmatrix},$$

then the candidate cells for $c_{10} = (2, 3)$ correspond to the nonnegative entries of J_{10} in cells $(i', 3)$ with $i' \geq 2$, namely when $(J_{10})_{3,3} = 1$ and $(J_{10})_{4,3} = 0$. So

$$C_{10} = \{(3, 3+1), (4, 3+0)\} = \{(3, 4), (4, 3)\}.$$

We must now modify backward slides as we did forward ones. Note that the following algorithm is to be applied *only* to pairs (T_k, J_k) that occur as output of the forward part of the algorithm. To avoid double subscripting, we let $(T', J') = (T_k, J_k)$.

SPN1 Pick $c \in C_k$ where $c_k = (i_0, j_0)$.

SPN2 **While** $c \neq c_k$ **do**

SPNa If $c = (i, j)$, then let c' be the cell of $\max\{T'_{i-1,j}, T'_{i,j-1}\}$ where we use 0 in place of $T'_{k,l}$ if $k < 0$ or $l < j_0$.

SPNb Exchange T'_c and $T'_{c'}$ and let $c := c'$. ∎

It is not clear that this procedure is well defined, since it might be possible, a priori, that we never have $c = c_k$. However, we will prove that c_k is always reached for any $c \in C_k$. We denote the resulting tableau by $j_c(T_k)$ and note that the condition in SPNa guarantees that since $T_k^{\leq c_k}$ was standard, $j_c(T_k)^{<c_k}$ will be standard as well. The sequence of cells encountered during the execution of j_c will be called the *reverse path* $r = r_c$ corresponding to c, and we let

$$\mathcal{R}_k = \{r_c \ : \ c \in C_k\}.$$

Going back to the previous example, we have

$$j_{(3,4)}(T_{10}) = \begin{matrix} 13 & 18 & 10 & 1 & 4 \\ 17 & 15 & 7 & 3 & 5 \\ 14 & 16 & 2 & 6 & 11 \\ 19 & 12 & 8 & 9 & \end{matrix} \ , \quad r_{(3,4)} = ((3,4), (3,3), (2,3)\};$$

$$j_{(4,3)}(T_{10}) = \begin{matrix} 13 & 18 & 10 & 1 & 4 \\ 17 & 15 & 8 & 3 & 5 \\ 14 & 16 & 2 & 7 & 11 \\ 19 & 12 & 6 & 9 & \end{matrix} \ , \quad r_{(4,3)} = ((4,3), (3,3), (2,3)\}.$$

To finish our description of the inverse algorithm, we must decide which of the candidate cells to use for our modified backward slide. Define the *code* of any reverse path r by replacing the steps of the form (i,j), $(i-1,j)$ by N (for northward) and those of the form (i,j), $(i,j-1)$ by W (for westward), and then reading the sequence of N's and W's backwards. Continuing with our running example, the codes for $r_{(3,4)}$ and $r_{(4,3)}$ are NW and NN, respectively. Define a lexicographic order on codes, and thus on reverse paths, by letting $N < \emptyset < W$. (When comparing codes of different lengths, the shorter one is padded out with $\emptyset$'s.). So in our example $NW > NN$, and so $r_{(3,4)} > r_{(4,3)}$. Now given (T_k, J_k) and $c_k = (i,j)$, we find the largest element r' in $\mathcal{R}_k$ with corresponding candidate cell $c' = (i',j')$ and let

$$T_{k-1} = j_{c'}(T_k).$$

Furthermore, $J_{k-1} = J_k$ except for the values

$$(J_{k-1})_{h,j} = \begin{cases} (J_k)_{h-1,j} + 1 & \text{for } i < h \leq i', \\ 0 & \text{for } h = i. \end{cases} \tag{3.20}$$

Unlike the forward definition, it is not obvious that J_{k-1} is well defined, since we might have $(J_{k-1})_{h,j} > a_{h,j}$ for some h. We will see shortly that this cannot happen.

So our next order of business is to show that SPN1–2 and (3.20) are well defined. First, however, we must set up some definitions and notation. Note that we can use SPN1–2 to define a reverse path, r, for any candidate cell, even if it does not pass through $c_k = (i_0, j_0)$, by letting r terminate when it reaches the top of column j_0 (which SPN2a will force it to do if it does not reach c_k). To talk about the relative positions of cells and paths, say that cell (i,j) is *west* (respectively *weakly west*) of cell (i',j') if $i = i'$ and $j < j'$ (respectively, $j \leq j'$). Cell c is *west* of path p' if it is west of some cell $c' \in p'$. Finally, path p is *west* of path p' if, for each $c \in p$ that is in the same row as a cell of p', c is west of some $c' \in p'$. Similar definitions apply to north, weakly north, etc.

We are now ready to start the well-definedness proofs, which will be by reverse induction on k. For ease of notation, let $T' = T_k$, $T'' = T_{k-1}$, $\mathcal{R}' = \mathcal{R}_k$, $\mathcal{R}'' = \mathcal{R}_{k-1}$, and so forth, and let $c_k = (i_0, j_0)$. The following lemma is fundamental for all that follows.

Lemma 3.10.4 *Suppose that all reverse paths in $\mathcal{R}'$ go through (i_0, j_0). Then r_0' with initial cell $(i_0', j_0') \in \mathcal{C}'$ is the largest reverse path in $\mathcal{R}'$ if and only if any initial cell $(i',j') \in \mathcal{C}'$ of $r' \in \mathcal{R}'$, $r' \neq r_0'$, satisfies*

R1 $i_0 \leq i' \leq i_0'$ *and* (i',j') *is west and weakly south of* r_0', *or*

R2 $i' > i_0'$ *and* r' *enters row* i_0' *weakly west of* r_0'.

When reading the proof of this lemma, the reader may find it useful to refer to the following diagram, which gives an example where $\lambda = (14^3, 13^4, 12)$, $(i_0, j_0) = (2,2)$, $(i_0', j_0') = (7,11)$, the lines indicate the steps of r_0', black dots are for possible cells in $\mathcal{C}'$, and white dots are for all other cells.

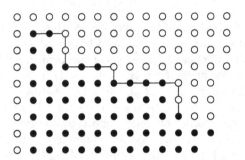

Proof. Since both r' and r'_0 end up at (i_0, j_0), they must intersect somewhere and coincide after their intersection. For the forward implication, assume that neither R1 nor R2 hold. Then the only other possibilities force r' to join r'_0 with a W step or start on r'_0 after an N step of r'_0. In either case $r' > r'_0$, a contradiction.

For the other direction, if r' satisfies R1 or R2, then either $r' \subset r'_0$ and r'_0 joins r' with a W step, or $r' \not\subseteq r'_0$ and r' joins r'_0 with an N step. This gives $r' < r'_0$. ∎

For our induction step we will need another lemma.

Lemma 3.10.5 *If r'_0 goes through (i_0, j_0) and is north of some cell of a reverse path $r'' \in \mathcal{R}''$, then r'' goes through $(i_0 + 1, j_0)$.*

Proof. I claim that r'_0 is north of every cell of r'' after the one given in the hypothesis of the lemma. This forces r'' through (i_0+1, j_0) as r' goes through (i_0, j_0). If the claim is false, then let (i_1, j_1) be the first cell of r'' after the given one that is northmost on r'_0 in its column. So the previous cell of r'' must have been $(i_1 + 1, j_1)$. But, by construction of r'_0,

$$T'_{i_1, j_1} = T''_{i_1, j_1 - 1} < T''_{i_1 + 1, j_1 - 1} = T'_{i_1 + 1, j_1 - 1}.$$

So r'' should have moved from (i_1+1, j_1) to (i_1+1, j_1-1) rather than (i_1, j_1), a contradiction. ∎

We will now simultaneously resolve both of our concerns about the inverse algorithm being well defined.

Proposition 3.10.6 *For all k, the hook tableau J_k is well defined and all reverse paths go through $c_k = (i_0, j_0)$ in T_k.*

Proof. The proposition is true for $i_0 = 1$ by definition of the algorithm. So by induction, we can assume that the statement is true for J' and T' and prove it for J'' and T''. By Lemma 3.10.4 and (3.20), we see that if $r'' \in R''$ starts at (i'', j'') with $i_0 \le i'' \le i'_0$, then (i'', j'') is south and weakly west of r'_0 as in the following figure (which the reader should compare with the previous one).

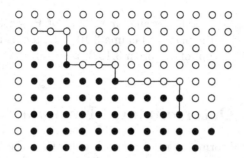

So, in particular, $(i'', j'') \in \lambda$ and J'' is well defined. Also, by Lemma 3.10.5, these r'' must go through $(i_0 + 1, j_0)$. If $i'' > i'_0$, let r' be the reverse path in T' starting at (i'', j''). Then r'' and r' must agree up to and including the first cell on r' in a column weakly west of column j'_0. Since this cell is south of r'_0, we are done by Lemma 3.10.5. ∎

It remains to show that the two maps are inverses of each other. Actually, it suffices to show that

1. if the pair (T'', J'') is derived from $(P, 0)$ by $k - 1$ applications of NPS, then applying NPS followed by SPN is the identity map, and

2. if the pair (T', J') is derived from (T, J) by $n - k$ applications of SPN, then applying SPN followed by NPS is the identity map.

The second statement is true because modified forward and backward slides are step-by-step inverses, and there is no question about which cell to use to undo the backward slide.

For the first one, we must show that if (T', J') is the pair after the last application of NPS, then the path p'_0 for the forward slide $j^{c_k}(T'')$ is just the largest reverse path $r'_0 \in \mathcal{R}'$ read backwards. We first need a lemma whose statement resembles that of Lemma 3.10.4.

Lemma 3.10.7 *Suppose that p'_0 ends at (i'_0, j'_0) and does not consist solely of E (east) steps. Then the initial cell $(i'', j'') \in C''$ of any $r'' \in \mathcal{R}''$ satisfies*

1. *$i_0 + 1 \leq i'' \leq i'_0$ and (i'', j'') is south and weakly west of p'_0, or*

2. *$i'' > i'_0$ and r'' enters row i'_0 weakly west of p'_0. ∎*

The proof of this result is by contradiction. It is also very similar to that of Lemma 3.10.5, and so is left to the reader.

By Lemma 3.10.4, to show that p'_0 backwards is the greatest element of $\mathcal{R}'$, it suffices to show the following.

Lemma 3.10.8 *The initial cell $(i', j') \in C'$ of any $r' \in \mathcal{R}'$ other than the reverse path of p'_0 satisfies*

1. *$i_0 \leq i' \leq i'_0$ and (i', j') is west and weakly south of p'_0, or*

2. *$i' > i'_0$ and r' enters row i'_0 weakly west of p'_0. ∎*

The proof of this last lemma is similar to that of Proposition 3.10.6, with Lemma 3.10.7 taking the place of Lemma 3.10.4. It, too, is left to the reader. This completes the proof of Theorem 3.10.3. ∎

3.11   The Determinantal Formula

The determinantal formula for f^λ is much older than the hook formula. In fact, the former was known to Frobenius [Fro 00, Fro 03] and Young [You 02]. In the following theorem we set $1/r! = 0$ if $r < 0$.

Theorem 3.11.1 (Determinantal Formula) *If* $(\lambda_1, \lambda_2, \ldots, \lambda_l) \vdash n$, *then*

$$f^\lambda = n! \left| \frac{1}{(\lambda_i - i + j)!} \right|,$$

where the determinant is l by l.

To remember the denominators in the determinant, note that the parts of λ occur along the main diagonal. The other entries in a given row are found by decreasing or increasing the number inside the factorial by 1 for every step taken to the left or right, respectively. Applying this result to $\lambda = (2, 2, 1)$, we get

$$f^{(2,2,1)} = 5! \cdot \begin{vmatrix} 1/2! & 1/3! & 1/4! \\ 1/1! & 1/2! & 1/3! \\ 0 & 1/0! & 1/1! \end{vmatrix} = 5,$$

which agrees with our computation in Section 3.10.

Proof (of the determinantal formula). It suffices to show that the determinant yields the denominator of the hook formula. From our definitions $\lambda_i + l = h_{i,1} + i$, so

$$\left| \frac{1}{(\lambda_i - i + j)!} \right| = \left| \frac{1}{(h_{i,1} - l + j)!} \right|.$$

Since every row of this determinant is of the form

$$\left[\frac{1}{(h - l + 1)!} \cdots \frac{1}{(h - 2)!} \frac{1}{(h - 1)!} \frac{1}{h!} \right]$$

and we will use only elementary column operations, we will display only the first row in what follows:

$$\left| \frac{1}{(h_{i,1} - l + j)!} \right| = \begin{vmatrix} \frac{1}{(h_{1,1}-l+1)!} & \cdots & \frac{1}{(h_{1,1}-2)!} & \frac{1}{(h_{1,1}-1)!} & \frac{1}{h_{1,1}!} \\ & & \vdots & & \end{vmatrix}$$

$$= \prod_{i=1}^{l} \frac{1}{h_{i,1}!} \cdot \begin{vmatrix} h_{1,1}(h_{1,1}-1)\cdots(h_{1,1}-l+2) & \cdots & h_{1,1}(h_{1,1}-1) & h_{1,1} & 1 \\ & \vdots & & & \end{vmatrix}$$

$$= \prod_{i=1}^{l} \frac{1}{h_{i,1}!} \cdot \begin{vmatrix} h_{1,1}(h_{1,1}-1)\cdots(h_{1,1}-l+2) & \cdots & h_{1,1}(h_{1,1}-1) & h_{1,1}-1 & 1 \\ & \vdots & & & \end{vmatrix}$$

$$= \prod_{i=1}^{l} \frac{1}{h_{i,1}!} \cdot \begin{vmatrix} h_{1,1}(h_{1,1}-1)\cdots(h_{1,1}-l+2) & \cdots & (h_{1,1}-1)(h_{1,1}-2) & h_{1,1}-1 & 1 \\ & \vdots & & & \end{vmatrix}$$

$$= \prod_{i=1}^{l} \frac{1}{h_{i,1}!} \cdot \begin{vmatrix} (h_{1,1}-1)(h_{1,1}-2)\cdots(h_{1,1}-l+1) & \cdots & (h_{1,1}-1)(h_{1,1}-2) & h_{1,1}-1 & 1 \\ & \vdots & & & \end{vmatrix}$$

$$= \prod_{i=1}^{l} \frac{1}{h_{i,1}} \cdot \begin{vmatrix} \frac{1}{(h_{1,1}-l)!} & \cdots & \frac{1}{(h_{1,1}-3)!} & \frac{1}{(h_{1,1}-2)!} & \frac{1}{(h_{1,1}-1)!} \\ & \vdots & & & \end{vmatrix}.$$

But by induction on n, this last determinant is just $1/\prod_{v \in \overline{\lambda}} h_v$, where

$$\begin{aligned} \overline{\lambda} &= (\lambda_1 - 1, \lambda_2 - 1, \ldots, \lambda_l - 1) \\ &= \lambda \text{ with its first column removed.} \end{aligned}$$

Note that $\overline{\lambda}$ may no longer have l rows, even though the determinant is still $l \times l$. However, induction can still apply, since the portion of the determinant corresponding to rows of $\overline{\lambda}$ of length zero is upper triangular with ones on the diagonal. Now $\overline{\lambda}$ contains all the hooklengths $h_{i,j}$ in λ for $i \geq 2$. Thus

$$\prod_{i=1}^{l} \frac{1}{h_{i,1}} \cdot \prod_{v \in \overline{\lambda}} \frac{1}{h_v} = \prod_{v \in \lambda} \frac{1}{h_v}$$

as desired. ∎

Notice that the same argument in reverse can be used to prove the hook formula for f^λ from the determinantal one. This was Frame, Robinson, and Thrall's original method of proof [FRT 54, pages 317–318], except that they started from a slightly different version of the Frobenius-Young formula (see Exercise 20).

3.12   Exercises

1. Prove Lemma 3.2.1.

2. Let P be a partial tableau with $x, y \notin P$. Suppose the paths of insertion for $r_x(P) = P'$ and $r_y(P') = P''$ are

$$(1, j_1), \ldots, (r, j_r) \quad \text{and} \quad (1, j_1'), \ldots, (r', j_{r'}'),$$

respectively. Show the following.

   (a) If $x < y$, then $j_i < j_i'$ for all $i \leq r'$ and $r' \leq r$; i.e., the path for r_y lies strictly to the right of the path for r_x.

   (b) If $x > y$, then $j_i \geq j_i'$ for all $i \leq r$ and $r' > r$; i.e., the path for r_y lies weakly to the left of the path for r_x.

3. (a) If the Robinson-Schensted algorithm is used only to find the length of the longest increasing subsequence of $\pi \in S_n$, find its computational complexity in the worst case.

   (b) Use the algorithm to find all increasing subsequences of π of maximum length.

   (c) Use Viennot's construction to find all increasing subsequences of π of maximum length.

4. (a) Use the results of this chapter to prove the Erdős-Szekeres theorem: Given any $\pi \in S_{nm+1}$, then π contains either an increasing subsequence of length $n + 1$ or a decreasing subsequence of length $m + 1$.

   (b) This theorem can be made into a game as follows. Player A picks $x_1 \in S = \{1, 2, \ldots, nm+1\}$. Then player B picks $x_2 \in S$ with $x_2 \neq x_1$. The players continue to alternate picking distinct elements of S until the sequence $x_1 x_2 \ldots$ contains an increasing subsequence of length $n + 1$ or a decreasing subsequence of length $m + 1$. In the achievement (respectively, avoidance) version of the game, the last player to move wins (respectively, loses). With m arbitrary, find winning strategies for $n \leq 2$ in the achievement game and for $n \leq 1$ in the avoidance game. It is an open problem to find a strategy in general.

5. Prove Lemma 3.4.5 and describe the Q-tableau of π_P.

6. Let $\pi = x_1 \ldots x_n$ be a partial permutation. The *Greene invariant of* π is the sequence of increasing subsequence lengths

$$i(\pi) = (i_1(\pi), i_2(\pi), \ldots, i_n(\pi)).$$

If $\pi \in S_n$, then let

$$\pi^{(j)} = x_1 \ldots x_j$$

and

$$\pi_{(j)} = \text{ the subsequence of } \pi \text{ containing all elements } \leq j.$$

(a) Show that $P(\pi) = P(\sigma)$ if and only if $i(\pi_{(j)}) = i(\sigma_{(j)})$ for all j by a direct argument.

(b) Show that $Q(\pi) = Q(\sigma)$ if and only if $i(\pi^{(j)}) = i(\sigma^{(j)})$ for all j by a direct argument.

(c) Show that (a) and (b) are equivalent statements.

7. Recall that an involution is a map π from a set to itself such that π^2 is the identity. Prove the following facts about involutions.

(a) A permutation π is an involution if and only if $P(\pi) = Q(\pi)$. Thus there is a bijection between involutions in S_n and standard tableaux with n elements.

(b) The number of fixedpoints in an involution π is the number of columns of odd length in $P(\pi)$.

(c) [Frobenius-Schur] The number of involutions in S_n is given by $\sum_{\lambda \vdash n} f^\lambda$.

(d) We have
$$1 \cdot 3 \cdot 5 \cdots (2n-1) = \sum_{\substack{\lambda \vdash 2n \\ \lambda' \text{ even}}} f^\lambda,$$

where λ' even means that the conjugate of λ has only even parts.

8. Let $\pi = x_1 x_2 \ldots x_n$ be any sequence of positive integers, possibly with repetitions. Define the *P-tableau of π*, $P(\pi)$, to be
$$r_{x_n} r_{x_{n-1}} \cdots r_{x_1}(\emptyset),$$
where the row insertion operator is defined as usual.

(a) Show that $P(\pi)$ is a semistandard tableau.

(b) Find and prove the semistandard analogues of each of the following results in the text. Some of the definitions may have to be modified. In each case you should try to obtain the new result as a corollary of the old one rather than rewriting the old proof to account for repetitions.

   i. Theorem 3.4.3.
   ii. Theorem 3.5.3.
   iii. Theorem 3.7.7.

9. Let P be a standard shifted tableau. Let P^2 denote the left-justified semistandard tableau gotten by pasting together P and P^t, as was done for shapes in Exercise 21.

Also define the insertion operation $i_x(P)$ by row inserting x into P until an element comes to rest at the end of a row (and insertion terminates) or a diagonal element in cell (d, d) is displaced. In the latter case $P_{d,d}$ bumps into *column $d+1$*, and bumping continues by columns until some element comes to rest at the end of a column.

  (a) Show that $(i_x P)^2 = c_x r_x (P^2)$, where the column insertion operator is modified so that x displaces the element of the first column greater than *or equal to* x, and so on.

  (b) Find and prove the shifted analogues of each of the following results in the text. Some of the definitions may have to be modified. In each case you should try to obtain the new result as a corollary of the old one rather than rewriting the old proof to account for the shifted tableaux.

     i. Theorem 3.4.3.

    ii. Theorem 3.5.3.

   iii. Theorem 3.7.7.

10. Reprove Proposition 3.8.1 using the Robinson-Schensted algorithm.

11. Consider the dihedral group D_4. Then $g \in D_4$ acts on $\pi \in S_n$ by applying the symmetry of the square corresponding to g to the permutation matrix corresponding to π. For every $g \in D_4$, describe $P(g\pi)$ and $Q(g\pi)$ in terms of $P(\pi)$ and $Q(\pi)$. Be sure your description is as simple as possible.

12. Let A be a partially ordered set. A *lower order ideal of P* is $L \subseteq P$ such that $x \in L$ implies that $y \in L$ for all $y \leq x$. *Upper order ideals* U are defined by the reverse inequality. If $P = L \uplus U$, where L and U are lower and upper order ideals, respectively, then define analogues of the jeu de taquin and vacating tableau on natural labelings of P (see Definition 4.2.5) such that

$$j^L(U) = v_U(L) \text{ and } v^L(U) = j_U(L).$$

Hint: Show that

$$j^L(U) \uplus v^L(U) = j_U(L) \uplus v_U(L)$$

by expressing each slide as a composition of operators that switch elements i and $i+1$ in a labeling as long as it remains natural.

13. Suppose $\pi = x_1 x_2 \ldots x_n$ is a permutation such that $P = P(\pi)$ has rectangular shape. Let the *complement of π* be

$$\pi^c = y_1 y_2 \ldots y_n,$$

where $y_i = n + 1 - x_i$ for all i. Also define the *complement of a rectangular standard tableau P* with n entries to be the array obtained by replacing $P_{i,j}$ with $n + 1 - P_{i,j}$ for all (i, j) and then rotating the result $180°$. Show that

$$P(\pi^c) = (P^c)^t.$$

14. The *reverse delta operator*, Δ', is the same as the delta operator of Definition 3.9.1, with m being replaced by the maximal element of Q and j^c replaced by j_c. The *reverse evacuation tableau*, $\mathrm{ev}'\, Q$, is the vacating tableau for the sequence

$$Q,\ \Delta' Q,\ \Delta'^2 Q,\ \ldots,\ \Delta'^n Q.$$

Prove the following.

(a) Evacuation and reverse evacuation are involutions, i.e.,

$$\mathrm{ev}\,\mathrm{ev}\, Q = \mathrm{ev}'\,\mathrm{ev}'\, Q = Q.$$

(b) If Q has rectangular shape, then

$$\mathrm{ev}'\,\mathrm{ev}\, Q = Q.$$

15. Prove Lemma 3.10.7.

16. Prove Lemma 3.10.8.

17. Fix a partition $\lambda \vdash n$ and consider the following algorithm due to Greene, Nijenhuis, and Wilf.

GNW1 Pick a node $v \in \lambda$ with probability $1/n$.
GNW2 **While** v is not an inner corner **do**
   GNWa Pick a node $\bar{v} \in H_v - \{v\}$ with probability $1/(h_v - 1)$.
   GNWb $v := \bar{v}$.
GNW3 Give the label n to the corner cell v that you have reached.
GNW4 Go back to step GNW1 with $\lambda := \lambda - \{v\}$ and $n := n - 1$, repeating this outer loop until all cells of λ are labeled.

The sequence of nodes generated by one pass through GNW1–3 is called a *trial*. In this exercise you will show that GNW1–4 produces any given standard λ-tableau P with probability

$$\mathrm{prob}(P) = \frac{\prod_{v \in \lambda} h_v}{n!}. \tag{3.21}$$

(a) Show that the algorithm always terminates and produces a standard tableau.

(b) Let (α, β) be the cell of P containing n and let $\mathrm{prob}(\alpha, \beta)$ be the probability that the first trial ends there. Prove that (3.21) follows by induction from the formula

$$\mathrm{prob}(\alpha, \beta) = \frac{1}{n} \prod_{i=1}^{\alpha-1} \left(1 + \frac{1}{h_{i,\beta} - 1}\right) \prod_{j=1}^{\beta-1} \left(1 + \frac{1}{h_{\alpha,j} - 1}\right). \tag{3.22}$$

(c) Given a trial ending at (α, β), we let the *horizontal projection of the trial* be

$$I = \{i \neq \alpha \; : \; v = (i, j) \text{ for some } v \text{ on the trial}\}.$$

The *vertical projection* is defined analogously. Let $\mathrm{prob}_{I,J}(\alpha, \beta)$ denote the sum of the probabilities of all trials terminating at (α, β) with horizontal projection I and vertical projection J. Show that

$$\mathrm{prob}_{I,J}(\alpha, \beta) = \frac{1}{n} \prod_{i \in I} \frac{1}{h_{i,\beta} - 1} \prod_{j \in J} \frac{1}{h_{\alpha,j} - 1}$$

and use this to prove (3.22). (Hint: Induct on $|I \cup J|$.)

(d) Give an alternative proof of the hook formula based on this exercise.

18. A poset τ is a *rooted tree* if it has a unique minimal element and its Hasse diagram has no cycles. If $v \in \tau$, then define its *hook* to be

$$H_v = \{w \in \tau \; : \; w \geq v\}$$

with corresponding *hooklength* $h_v = |H_v|$.

(a) Show that

$$\sum_\tau (f^\tau)^2 = n!$$

where the sum is over all τ with n nodes.

(b) Show in three ways (inductively, probabilistically, and using jeu de taquin) that if τ has n nodes, then the number of natural labelings of τ (see Definition 4.2.5) is

$$f^\tau = \frac{n!}{\prod_{v \in \tau} h_v}.$$

19. A *two-person ballot sequence* is a permutation $\pi = x_1 x_2 \ldots x_{2n}$ of n ones and n twos such that, for any prefix $\pi_k = x_1 x_2 \ldots x_k$, the number of ones in π_k is at least as great as the number of twos. The nth *Catalan number*, C_n, is the number of such sequences.

(a) Prove the recurrence

$$C_{n+1} = C_n C_0 + C_{n-1} C_1 + \cdots + C_0 C_n$$

for $n \geq 0$.

(b) The Catalan numbers also count the following sets. Show this in two ways: by verifying that the recurrence is satisfied and by giving a bijection with a set of objects already known to be counted by the C_n.

      i. Standard tableaux of shape (n, n).

     ii. Permutations $\pi \in \mathcal{S}_n$ with longest decreasing subsequence of length at most two.

   iii. Sequences of positive integers

$$1 \leq a_1 \leq a_2 \leq \cdots \leq a_n$$

such that $a_i \leq i$ for all i.

   iv. Binary trees on n nodes.

    v. Pairings of $2n$ labeled points on a circle with chords that do not cross.

   vi. Triangulations of a convex $(n + 2)$-gon by diagonals.

  vii. Lattice paths of the type considered in Section 4.5 from $(0,0)$ to (n, n) that stay weakly below the diagonal $y = x$.

 viii. Noncrossing partitions of $\{1, 2, \ldots, n\}$, i.e., ways of writing

$$\{1, 2, \ldots, n\} = B_1 \uplus B_2 \uplus \cdots \uplus B_k$$

such that $a, c \in B_i$ and $b, d \in B_j$ with $a < b < c < d$ implies $i = j$.

(c) Prove, by using results of this chapter, that

$$C_n = \frac{1}{n+1}\binom{2n}{n}.$$

20. If $\lambda = (\lambda_1, \lambda_2, \ldots, \lambda_l) \vdash n$, then derive the following formulae

$$
\begin{aligned}
f^\lambda &= n!\frac{\prod_{i<j}(\lambda_i - \lambda_j - i + j)}{\prod_i (\lambda_i - i + l)!} \\
&= n!\frac{\prod_{i<j}(h_{i,1} - h_{j,1})}{\prod_i h_{i,1}!}.
\end{aligned}
$$

21. A partition $\lambda = (\lambda_1, \lambda_2, \ldots, \lambda_l) \vdash n$ is *strict* if $\lambda_1 > \lambda_2 > \cdots > \lambda_l$. Given any strict partition, the associated *shifted shape* λ^* indents row i of the normal shape so that it starts on the diagonal square (i, i). For example, the shifted shape of $\lambda = (4, 3, 1)$ is

$$\lambda^* = \quad \begin{matrix} \bullet & \bullet & \bullet & \bullet \\ & \bullet & \bullet & \bullet \\ & & \bullet \end{matrix} \; .$$

The *shifted hook* of $(i, j) \in \lambda^*$ is

$$H^*_{i,j} = \{(i, j') \; : \; j' \geq j\} \cup \{(i', j) \; : \; i' \geq i\} \cup \{(j+1, j') \; : \; j' \geq j+1\}$$

with *shifted hooklength* $h^*_{i,j} = |H^*_{i,j}|$. For λ^* as before, the shifted hooklengths are

$$
\begin{matrix}
7 & 5 & 4 & 2 \\
  & 4 & 3 & 1 \; . \\
  &   & 1
\end{matrix}
$$

(a) Show that the $h^*_{i,j}$ can be obtained as ordinary hooklengths in a left-justified diagram obtained by pasting together λ^* with its transpose.

(b) Defining standard shifted tableaux in the obvious way, show that the number of such arrays is given by

$$
\begin{aligned}
f^{\lambda^*} &= n! \frac{\prod_{i<j}(\lambda_i - \lambda_j)}{\prod_i \lambda_i! \prod_{i<j}(\lambda_i + \lambda_j)} \\
&= \frac{n!}{\prod_{i<j}(\lambda_i + \lambda_j)} \det \frac{1}{(\lambda_i - l + j)!} \\
&= \frac{n!}{\prod_{(i,j)\in\lambda^*} h^*_{i,j}}.
\end{aligned}
$$

Chapter 4

Symmetric Functions

We have seen how some results about representations of $\mathcal{S}_n$ can be proved either by using general facts from representation theory or combinatorially. There is a third approach using symmetric functions, which is our focus in this chapter.

After giving some general background on formal power series, we derive the hook generating function for semistandard tableaux (generalizing the hook formula, Theorem 3.10.2). The method of proof is a beautiful algorithmic bijection due to Hillman and Grassl [H-G 76].

Next, the symmetric functions themselves are introduced along with the all-important Schur functions, s_λ. The Jacobi-Trudi determinants give alternative expressions for s_λ analogous to the determinantal form for f^λ and lattice-path techniques can be used to provide a combinatorial proof. Other definitions of the Schur function as a quotient of alternates or as the cycle indicator for the irreducible characters of $\mathcal{S}_n$ are presented. The latter brings in the characteristic map, which is an isomorphism between the algebra of symmetric functions and the algebra of class functions on the symmetric group (which has the irreducible characters as a basis). Knuth's generalization of the Robinson-Schensted map [Knu 70] completes our survey of generating function analogues for results from the previous chapter.

We end by coming full circle with two applications of symmetric functions to representation theory. The first is the Littlewood-Richardson rule [L-R 34], which decomposes a tensor product into irreducibles by looking at the corresponding product of Schur functions. We present a proof based on the jeu de taquin and dual equivalence. The second is a theorem of Murnaghan-Nakayama [Nak 40, Mur 37] which gives an algorithm for computing the irreducible characters. Its proof involves much of the machinery that has been introduced previously.

Those who would like a more extensive and algebraic treatment of symmetric functions should consult Macdonald's book [Mac 79].

4.1   Introduction to Generating Functions

We start with the most basic definition.

Definition 4.1.1 Given a sequence $(a_n)_{n \geq 0} = a_0, a_1, a_2, \ldots$ of complex numbers, the corresponding *generating function* is the power series

$$f(x) = \sum_{n \geq 0} a_n x^n.$$

If the a_n enumerate some set of combinatorial objects, then we say that $f(x)$ is the generating function for those objects. We also write

$$[x^n]f(x) = \text{ the coefficient of } x^n \text{ in } f(x) = a_n. \blacksquare$$

For example, if

$$a_n = \text{ the number of } n\text{-element subsets of } \{1, 2, 3\} \stackrel{\text{def}}{=} \binom{3}{n}$$

then

$$a_0 = 1, \ a_1 = 3, \ a_2 = 3, \ a_3 = 1, \text{ and } a_n = 0 \text{ for } n \geq 4,$$

so

$$f(x) = \sum_{n \geq 0} \binom{3}{n} x^n = 1 + 3x + 3x^2 + x^3 = (1 + x)^3$$

is the generating function for subsets of $\{1, 2, 3\}$. Equivalently,

$$[x^n](1 + x)^3 = \binom{3}{n}.$$

It may seem surprising, but to obtain information about a sequence it is often easier to manipulate its generating function. In particular, as we will see shortly, sometimes there is no known simple expression for a_n and yet $f(x)$ is easy to compute. Extensive discussions of generating function techniques can be found in the texts of Goulden and Jackson [G-J 83], Stanley [Stn 97, Stn 99], and Wilf [Wil 90].

Note that all our power series are members of the *formal power series ring*

$$\mathbb{C}[[x]] = \{ \sum_{n \geq 0} a_n x^n \ : \ a_n \in \mathbb{C} \text{ for all } n \},$$

which is a ring under the usual operations of addition and multiplication. The adjective *formal* refers to the fact that convergence questions are immaterial, since we will never substitute a value for x. The variable and its powers are merely being used to keep track of the coefficients.

We need techniques for deriving generating functions. The basic counting rules for sets state that the English words *or* and *and* are the equivalent of the mathematical operations $+$ and $\times$. Formally, we have the following.

Proposition 4.1.2 *Let S and T be finite sets.*

1. *If $S \cap T = \emptyset$, then*

$$|S \uplus T| = |S| + |T|.$$

2. *If S and T are arbitrary, then*

$$|S \times T| = |S| \cdot |T|. \blacksquare$$

This result has an analogue for generating functions (see Proposition 4.1.6). First, however, let us see how these rules can be used informally to compute a few examples.

A basic method for finding the generating function for a given sequence $(a_n)_{n \geq 0}$ is as follows:

1. Find a set S with a parameter such that the number of elements of S whose parameter equals n is a_n.

2. Express the elements of S in terms of *or, and,* and the parameter.

3. Translate this expression into a generating function using $+$, $\times$, and x^n.

In our previous example, $a_n = \binom{3}{n}$, so we can take

$$S = \text{ all subsets } T \text{ of } \{1, 2, 3\}.$$

The parameter that will produce the sequence is

$$n = n(T) = \text{ the number of elements in } T.$$

Now we can express any such subset as

$$T = (1 \notin T \text{ or } 1 \in T) \quad \text{and} \quad (2 \notin T \text{ or } 2 \in T) \quad \text{and} \quad (3 \notin T \text{ or } 3 \in T).$$

Finally, translate this expression into a generating function. Remember that the n in x^n is the number of elements in T, so the statements $i \notin T$ and $i \in T$ become x^0 and x^1, respectively. So

$$f(x) = (x^0 + x^1) \cdot (x^0 + x^1) \cdot (x^0 + x^1) = (1 + x)^3$$

as before.

For a more substantial example of the method, let us find the generating function

$$\sum_{n \geq 0} p(n) x^n,$$

where $p(n)$ is the number of partitions of n. Here, S is all partitions $\lambda = (1^{m_1}, 2^{m_2}, \ldots)$ and $n = |\lambda|$ is the sum of the parts. We have

$$
\begin{aligned}
\lambda \;=\; & (1^0 \in \lambda \text{ or } 1^1 \in \lambda \text{ or } 1^2 \in \lambda \text{ or } \cdots) \\
& \text{and} \quad (2^0 \in \lambda \text{ or } 2^1 \in \lambda \text{ or } 2^2 \in \lambda \text{ or } \cdots) \\
& \text{and} \quad (3^0 \in \lambda \text{ or } 3^1 \in \lambda \text{ or } 3^2 \in \lambda \text{ or } \cdots) \quad \text{and} \cdots,
\end{aligned}
$$

which translates as

$$f(x) = (x^0+x^1+x^{1+1}+\cdots)(x^0+x^2+x^{2+2}+\cdots)(x^0+x^3+x^{3+3}+\cdots)\cdots. \quad (4.1)$$

Thus we have proved a famous theorem of Euler [Eul 48].

Theorem 4.1.3 *The generating function for partitions is*

$$\sum_{n\geq 0} p(n)x^n = \frac{1}{1-x}\frac{1}{1-x^2}\frac{1}{1-x^3}\cdots. \ \blacksquare \quad (4.2)$$

Several remarks about this result are in order. Despite the simplicity of this generating function, there is no known closed-form formula for $p(n)$ itself. (However, there is an expression for $p(n)$ as a sum due to Hardy, Ramanujan, and Rademacher; see Theorem 5.1 on page 69 of Andrews [And 76].) This illustrates the power of our approach.

Also, the reader should be suspicious of infinite products such as (4.2). Are they really well-defined elements of $\mathbb{C}[[x]]$? To see what can go wrong, try to find the coefficient of x in $\prod_{i=1}^{\infty}(1+x)$. To deal with this problem, we need some definitions. Consider $f(x), f_1(x), f_2(x), \ldots \in \mathbb{C}[[x]]$. Then we write $\prod_{i\geq 1} f_i(x) = f(x)$ and say that the product *converges* to $f(x)$ if, for every n,

$$[x^n]f(x) = [x^n]\prod_{i=1}^{N} f_i(x)$$

whenever N is sufficiently large. Of course, how large N needs to be depends on n.

A convenient condition for convergence is expressed in terms of the *degree* of $f(x) \in \mathbb{C}[[x]]$, where

$$\deg f(x) = \text{ smallest } n \text{ such that } x^n \text{ has nonzero coefficient in } f(x).$$

For example,

$$\deg(x^2 + x^3 + x^4 + \cdots) = 2.$$

The following proposition is not hard to prove and is left to the reader.

Proposition 4.1.4 *If $f_i(x) \in \mathbb{C}[[x]]$ for $i \geq 1$ and $\lim_{i\to\infty} \deg(f_i(x) - 1) = \infty$, then $\prod_{i\geq 1} f_i(x)$ converges.* $\blacksquare$

Note that this shows that the right-hand side of equation (4.2) makes sense, since there $\deg(f_i(x) - 1) = i$.

By carefully examining the derivation of Euler's result, we see that the term $1/(1-x^i)$ in the product counts the occurrences of i in the partition λ. Thus we can automatically construct other generating functions. For example, if we let

$$p_o(n) = \text{ the number of } \lambda \vdash n \text{ with all parts odd,}$$

then

$$\sum_{n \geq 0} p_o(n)x^n = \prod_{i \geq 1} \frac{1}{1 - x^{2i-1}}.$$

We can also keep track of the number of parts. To illustrate, consider

$$p_d(n) = \text{ the number of } \lambda \vdash n \text{ with all parts distinct,}$$

i.e., no part of λ appears more than once. Thus the only possibilities for a part i are $i^0 \in \lambda$ or $i^1 \in \lambda$. This amounts to cutting off the generating function in (4.1) after the first two terms, so

$$\sum_{n \geq 0} p_d(n)x^n = \prod_{i \geq 1}(1 + x^i).$$

As a final demonstration of the utility of generating functions, we use them to derive another theorem of Euler [Eul 48] .

Theorem 4.1.5 *For all n, $p_d(n) = p_o(n)$.*

Proof. It suffices to show that $p_d(n)$ and $p_o(n)$ have the same generating function. But

$$\prod_{i \geq 1}(1 + x^i) = \prod_{i \geq 1}(1 + x^i) \prod_{i \geq 1} \frac{1 - x^i}{1 - x^i}$$
$$= \prod_{i \geq 1} \frac{1 - x^{2i}}{1 - x^i}$$
$$= \prod_{i \geq 1} \frac{1}{1 - x^{2i-1}}. \blacksquare$$

It is high time to make more rigorous the steps used to derive generating functions. The crucial definition is as follows. Let S be a se. Then a *weighting of S* is a function

$$\text{wt} : S \to \mathbb{C}[[x]].$$

If $s \in S$, then we usually let $\text{wt } s = x^n$ for some n, what we were calling a parameter earlier. The associated *weight generating function* is

$$f_S(x) = \sum_{s \in S} \text{wt } s.$$

It is a well-defined element of $\mathbb{C}[[x]]$ as long as the number of $s \in S$ with $\deg \text{wt } s = n$ is finite for every n. To redo our example for the partition function, let S be the set of all partitions, and if $\lambda \in S$, then define

$$\text{wt } \lambda = x^{|\lambda|}.$$

This gives the weight generating function

$$
\begin{aligned}
f_S(x) &= \sum_{\lambda \in S} x^{|\lambda|} \\
&= \sum_{n \geq 0} \sum_{\lambda \vdash n} x^n \\
&= \sum_{n \geq 0} p(n) x^n,
\end{aligned}
$$

which is exactly what we wish to evaluate.

In order to manipulate weights, we need the corresponding and-or rules.

Proposition 4.1.6 *Let S and T be weighted sets.*

1. *If $S \cap T = \emptyset$, then*

$$
f_{S \uplus T}(x) = f_S(x) + f_T(x).
$$

2. *Let S and T be arbitrary and weight $S \times T$ by $\mathrm{wt}(s,t) = \mathrm{wt}\, s\, \mathrm{wt}\, t$. Then*

$$
f_{S \times T} = f_S(x) \cdot f_T(x).
$$

Proof. 1. If S and T do not intersect, then

$$
\begin{aligned}
f_{S \uplus T}(x) &= \sum_{s \in S \uplus T} \mathrm{wt}\, s \\
&= \sum_{s \in S} \mathrm{wt}\, s + \sum_{s \in T} \mathrm{wt}\, s \\
&= f_S(x) + f_T(x).
\end{aligned}
$$

2. For any two sets S, T, we have

$$
\begin{aligned}
f_{S \times T}(x) &= \sum_{(s,t) \in S \times T} \mathrm{wt}(s,t) \\
&= \sum_{\substack{s \in S \\ t \in T}} \mathrm{wt}\, s\, \mathrm{wt}\, t \\
&= \sum_{s \in S} \mathrm{wt}\, s \sum_{t \in T} \mathrm{wt}\, t \\
&= f_S(x) f_T(x). \qquad \blacksquare
\end{aligned}
$$

Under suitable convergence conditions, this result can be extended to infinite sums and products. Returning to the partition example:

$$
\begin{aligned}
S &= \{\lambda = (1^{m_1}, 2^{m_2}, \ldots) : m_i \geq 0\} \\
&= (1^0, 1^1, 1^2, \ldots) \amalg (2^0, 2^1, 2^2, \ldots) \amalg \cdots \\
&= (\{1^0\} \uplus \{1^1\} \uplus \{1^2\} \uplus \cdots) \amalg (\{2^0\} \uplus \{2^1\} \uplus \{2^2\} \uplus \cdots) \amalg \cdots,
\end{aligned}
$$

where $\amalg$ rather than $\times$ is being used, since one is allowed to take only a finite number of components i^{m_i} such that $m_i \neq 0$. So

$$
f_S(x) = (f_{\{1^0\}} + f_{\{1^1\}} + f_{\{1^2\}} + \cdots)(f_{\{2^0\}} + f_{\{2^1\}} + f_{\{2^2\}} + \cdots) \cdots.
$$

Since $f_{\{\lambda\}}(x) = \mathrm{wt}\, \lambda = x^{|\lambda|}$, we recover (4.1) as desired.

4.2   The Hillman-Grassl Algorithm

Just as we were able to enumerate standard λ-tableaux in Chapter 3, we wish to count the semistandard variety of given shape. However, since there are now an infinite number of such arrays, we have to use generating functions. One approach is to sum up the parts, as we do with partitions. This leads to a beautiful hook generating function that was first discovered by Stanley [Stn 71]. The combinatorial proof given next is due to Hillman and Grassl [H-G 76].

Fix a shape λ and let $ss_\lambda(n)$ be the number of semistandard λ-tableaux T such that $\sum_{(i,j)\in\lambda} T_{i,j} = n$. Since all entries in row i of T are at least of size i, we can replace each $T_{i,j}$ by $T_{i,j} - i$ to obtain a new type of tableau.

Definition 4.2.1 A *reverse plane partition of shape* λ, T, is an array obtained by replacing the nodes of λ by nonnegative integers so that the rows and columns are weakly increasing. If the entries in of T sum to n, we say that T is a *reverse plane partition of* n. Let

$rpp_\lambda(n) = $ the number of reverse plane partitions of n having shape λ. ∎

The use of the term *reverse* for plane partitions where the parts *increase* is a historical accident stemming from the fact that ordinary partitions are usually written in weakly decreasing order.

From the definitions, we clearly have

$$\sum_{n\geq 0} ss_\lambda(n)x^n = x^{m(\lambda)} \sum_{n\geq 0} rpp_\lambda(n)x^n,$$

where $m(\lambda) = \sum_{i\geq 1} i\lambda_i$. Thus it suffices to find the generating function for reverse plane partitions. Once again, the hooklengths come into play.

Theorem 4.2.2 *Fix a partition* λ. *Then*

$$\sum_{n\geq 0} rpp_\lambda(n)x^n = \prod_{(i,j)\in\lambda} \frac{1}{1 - x^{h_{i,j}}}.$$

Proof. By the discussion after Proposition 4.1.4, the coefficient of x^n in this product counts partitions of n, where each part is of the form $h_{i,j}$ for some $(i,j) \in \lambda$. (Note that the part $h_{i,j}$ is associated with the node $(i,j) \in \lambda$, so parts $h_{i,j}$ and $h_{k,l}$ are considered different if $(i,j) \neq (k,l)$ even if $h_{i,j} = h_{k,l}$ as integers.) To show that this coefficient equals the number of reverse plane partitions T of n, it suffices to find a bijection

$$T \longleftrightarrow (h_{i_1,j_1}, h_{i_2,j_2}, \ldots)$$

that is weight preserving, i.e.,

$$\sum_{(i,j)\in\lambda} T_{i,j} = \sum_k h_{i_k,j_k}.$$

"$T \to (h_{i_1,j_1}, h_{i_2,j_2}, \ldots)$" Given T, we will produce a sequence of reverse plane partitions

$$T = T_0, T_1, T_2, \ldots, T_f = \text{ tableau of zeros,}$$

where T_k will be obtained from T_{k-1} by subtracting 1 from all elements of a certain *path* of cells p_k in T_k. Since we will always have $|p_k| = h_{i_k,j_k}$ for some (i_k, j_k), this will ensure the weight-preserving condition.

Define the path $p = p_1$ in T inductively as follows.

HG1 Start p at (a, b), the most northeast cell of T containing a nonzero entry.

HG2 Continue by

$$\textbf{if } (i,j) \in p \textbf{ then} \begin{cases} (i, j-1) \in p & \text{if } T_{i,j-1} = T_{i,j}, \\ (i+1, j) \in p & \text{otherwise.} \end{cases}$$

In other words, move south unless forced to move west in order not to violate the weakly increasing condition along the rows (once the ones are subtracted).

HG3 Terminate p when the preceding induction rule fails. At this point we must be at the end of some column, say column c.

It is easy to see that after subtracting 1 from the elements in p, the array remains a reverse plane partition and the amount subtracted is $h_{a,c}$.

As an example, let

$$T = \begin{matrix} 1 & 2 & 2 & 2 \\ 3 & 3 & 3 \\ 3 \end{matrix} \quad .$$

Then $(a, b) = (1, 4)$, and the path p is indicated by the dotted cells in the following diagram:

After subtraction, we have

$$T_1 = \begin{matrix} 1 & 1 & 1 & 1 \\ 2 & 2 & 3 \\ 2 \end{matrix}$$

and $h_{i_1,j_1} = h_{1,1}$. To obtain the rest of the T_k, we iterate this process. The complete list for our example, together with the corresponding $h_{i,j}$, is

	1 2 2 2	1 1 1 1	0 0 0 0	0 0 0 0	0 0 0 0	0 0 0 0
T_k:	3 3 3 ,	2 2 3 ,	1 2 3 ,	1 2 2 ,	1 1 1 ,	0 0 0 ;
	3	2	1	1	1	0
h_{i_k,j_k}:		$h_{1,1}$,	$h_{1,1}$,	$h_{2,3}$,	$h_{2,2}$,	$h_{2,1}$.

Thus $T \to (h_{1,1}, h_{1,1}, h_{2,3}, h_{2,2}, h_{2,1})$.

"$(h_{i_1,j_1}, h_{i_2,j_2}, \ldots) \to T$" Given a partition of hooklengths, we must re-build the reverse plane partition. First, however, we must know in what order the hooklengths were removed.

Lemma 4.2.3 *In the decomposition of T into hooklengths, the hooklength $h_{i',j'}$ was removed before $h_{i'',j''}$ if and only if*

$$i'' > i', \quad or \quad i'' = i' \text{ and } j'' \leq j'. \tag{4.3}$$

Proof. Since (4.3) is a total order on the nodes of the shape, we need only prove the only-if direction. By transitivity, it suffices to consider the case where $h_{i'',j''}$ is removed directly after $h_{i',j'}$.

Let T' and T'' be the arrays from which $h_{i',j'}$ and $h_{i'',j''}$ were removed using paths p' and p'', respectively. By the choice of initial points and the fact that entries decrease in passing from T' to T'', we have $i'' \geq i'$.

If $i'' > i'$, we are done. Otherwise, $i'' = i'$ and p'' starts in a column weakly to the west of p'. We claim that in this case p'' can never pass through a node strictly to the east of a node of p', forcing $j'' \leq j'$. If not, then there is some $(s, t) \in p' \cap p''$ such that $(s, t-1) \in p'$ and $(s+1, t) \in p''$. But the fact that p' moved west implies $T'_{s,t} = T'_{s,t-1}$. Since this equality continues to hold in T'' after the ones have been subtracted, p'' is forced to move west as well, a contradiction. ∎

Returning to the construction of T, if we are given a partition of hook-lengths, then order them as in the lemma; suppose we get

$$(h_{i_1,j_1}, \ldots, h_{i_f,j_f}).$$

We then construct a sequence of tableaux, starting with the all-zero array,

$$T_f, \; T_{f-1}, \; \ldots, \; T_0 = T,$$

by adding back the h_{i_k,j_k} for $k = f, f-1, \ldots, 1$. To add $h_{a,c}$ to T, we construct a *reverse path* r along which to add ones.

GH1 Start r at the most southern node in column c.

GH2 Continue by

$$\textbf{if } (i,j) \in r \textbf{ then} \begin{cases} (i, j+1) \in r & \text{if } T_{i,j+1} = T_{i,j}, \\ (i-1, j) \in r & \text{otherwise.} \end{cases}$$

GH3 Terminate r when it passes through the eastmost node of the row a.

It is clear that this is a step-by-step inverse of the construction of the path p. However, it is not clear that r is well defined, i.e., that it must pass through the eastern end of row a. Thus to finish the proof of Theorem 4.2.2, it suffices to prove a last lemma.

Lemma 4.2.4 *If r_k is the reverse path for h_{i_k,j_k}, then $(i_k, \lambda_{i_k}) \in r_k$.*

Proof. Use reverse induction on k. The result is obvious when $k = f$ by the first alternative in step GH2.

For $k < f$, let $r' = r_k$ and $r'' = r_{k+1}$. Similarly, define $T', T'', h_{i',j'}$, and $h_{i'',j''}$. By our ordering of the hooklengths, $i' \leq i''$. If $i' < i''$, then row i' of T consists solely of zeros, and we are done as in the base case.

If $i' = i''$, then $j' \geq j''$. Thus p' starts weakly to the east of p''. By the same arguments as in Lemma 4.3, p stays to the east of p'. Since p' reaches the east end of row $i' = i$ by assumption, so must p. ∎

It is natural to ask whether there is any relation between the hook formula and the hook generating function. In fact, the former is a corollary of the latter if we appeal to some general results of Stanley [Stn 71] about poset partitions.

Definition 4.2.5 *Let $(A, \leq)$ be a partially ordered set. A reverse A-partition of m is an order-preserving map*

$$\alpha : A \to \{0, 1, 2, \ldots\}$$

such that $\sum_{v \in A} \alpha(v) = m$. If $|A| = n$, then an order-preserving bijection

$$\beta : A \to \{1, 2, \ldots, n\}$$

is called a natural labeling of A. ∎

To see the connection with tableaux, partially order the cells of λ in the natural way,
$$(i,j) \leq (i',j') \iff i \leq i' \quad \text{and} \quad j \leq j'.$$
Then a natural labeling of λ is just a standard λ-tableau, whereas a reverse λ-partition is a reverse plane partition of shape λ.

One of Stanley's theorems about poset partitions is the following.

Theorem 4.2.6 ([Stn 71]) *Let A be a poset with $|A| = n$. Then the generating function for reverse A-partitions is*

$$\frac{P(x)}{(1 - x)(1 - x^2) \cdots (1 - x^n)},$$

where $P(x)$ is a polynomial such that $P(1)$ is the number of natural labelings of A. ∎

In the case where $A = \lambda$, we can compare this result with Theorem 4.2.2 and obtain

$$\frac{P(x)}{(1 - x)(1 - x^2) \cdots (1 - x^n)} = \prod_{(i,j) \in \lambda} \frac{1}{1 - x^{h_{i,j}}}.$$

Thus

$$
\begin{aligned}
f^\lambda &= P(1) \\
&= \lim_{x \to 1} \frac{\prod_{k=1}^n (1 - x^k)}{\prod_{(i,j) \in \lambda}(1 - x^{h_{i,j}})} \\
&= \frac{n!}{\prod_{(i,j) \in \lambda} h_{i,j}}.
\end{aligned}
$$

4.3   The Ring of Symmetric Functions

In order to keep track of more information with our generating functions, we can use more than one variable. The ring of symmetric functions then arises as a set of power series invariant under the action of all the symmetric groups.

Let $\mathbf{x} = \{x_1, x_2, x_3, \ldots\}$ be an infinite set of variables and consider the formal power series ring $\mathbb{C}[[\mathbf{x}]]$. The monomial $x_{i_1}^{\lambda_1} x_{i_2}^{\lambda_2} \cdots x_{i_l}^{\lambda_l}$ is said to have *degree* n if $n = \sum_i \lambda_i$. We also say that $f(\mathbf{x}) \in \mathbb{C}[[\mathbf{x}]]$ is *homogeneous of degree* n if every monomial in $f(\mathbf{x})$ has degree n.

. For every n, there is a natural action of $\pi \in S_n$ on $f(\mathbf{x}) \in \mathbb{C}[[\mathbf{x}]]$, namely,

$$
\pi f(x_1, x_2, x_3, \ldots) = f(x_{\pi 1}, x_{\pi 2}, x_{\pi 3}, \ldots), \tag{4.4}
$$

where $\pi i = i$ for $i > n$. The simplest functions fixed by this action are gotten by symmetrizing a monomial.

Definition 4.3.1 Let $\lambda = (\lambda_1, \lambda_2, \ldots, \lambda_l)$ be a partition. The *monomial symmetric function corresponding to* λ is

$$
m_\lambda = m_\lambda(\mathbf{x}) = \sum x_{i_1}^{\lambda_1} x_{i_2}^{\lambda_2} \cdots x_{i_l}^{\lambda_l},
$$

where the sum is over all distinct monomials having exponents $\lambda_1, \ldots, \lambda_l$. ∎

For example,

$$
m_{(2,1)} = x_1^2 x_2 + x_1 x_2^2 + x_1^2 x_3 + x_1 x_3^2 + x_2^2 x_3 + x_2 x_3^2 + \cdots.
$$

Clearly, if $\lambda \vdash n$, then $m_\lambda(\mathbf{x})$ is homogeneous of degree n.

Now we can define the symmetric functions that interest us.

Definition 4.3.2 The *ring of symmetric functions* is

$$
\Lambda = \Lambda(\mathbf{x}) = \mathbb{C} m_\lambda,
$$

i.e., the vector space spanned by all the m_λ. ∎

Note that Λ is really a ring, not just a vector space, since it is closed under product. However, there are certain elements of $\mathbb{C}[[\mathbf{x}]]$ invariant under (4.4) that are not in Λ, such as $\prod_{i \geq 1}(1 + x_i)$, which cannot be written as a *finite* linear combination of m_λ.

We have the decomposition

$$\Lambda = \oplus_{n \geq 0} \Lambda^n,$$

where Λ^n is the space spanned by all m_λ of degree n. In fact, this is a *grading* of Λ, since

$$f \in \Lambda^n \quad \text{and} \quad g \in \Lambda^m \quad \text{implies} \quad fg \in \Lambda^{n+m}. \tag{4.5}$$

Since the m_λ are independent, we have the following result.

Proposition 4.3.3 *The space* Λ^n *has basis*

$$\{m_\lambda \ : \ \lambda \vdash n\}$$

and so has dimension $p(n)$, *the number of partitions of* n. ∎

There are several other bases for Λ^n that are of interest. To construct them, we need the following families of symmetric functions.

Definition 4.3.4 The *nth power sum symmetric function* is

$$p_n = m_{(n)} = \sum_{i \geq 1} x_i^n.$$

The *nth elementary symmetric function* is

$$e_n = m_{(1^n)} = \sum_{i_1 < \cdots < i_n} x_{i_1} \cdots x_{i_n}.$$

The *nth complete homogeneous symmetric function* is

$$h_n = \sum_{\lambda \vdash n} m_\lambda = \sum_{i_1 \leq \cdots \leq i_n} x_{i_1} \cdots x_{i_n} \ \blacksquare.$$

As examples, when $n = 3$,

$$
\begin{aligned}
p_3 &= x_1^3 + x_2^3 + x_3^3 + \cdots, \\
e_3 &= x_1 x_2 x_3 + x_1 x_2 x_4 + x_1 x_3 x_4 + x_2 x_3 x_4 + \cdots, \\
h_3 &= x_1^3 + x_2^3 + \cdots + x_1^2 x_2 + x_1 x_2^2 + \cdots + x_1 x_2 x_3 + x_1 x_2 x_4 + \cdots.
\end{aligned}
$$

The elementary function e_n is just the sum of all square-free monomials of degree n. As such, it can be considered as a weight generating function for partitions with n distinct parts. Specifically, let

$$S = \{\lambda \ : \ l(\lambda) = n\},$$

where $l(\lambda)$ is the number of parts of λ, known as its *length*. If $\lambda = (\lambda_1 > \lambda_2 > \cdots > \lambda_n)$, we use the weight

$$\text{wt } \lambda = x_{\lambda_1} x_{\lambda_2} \cdots x_{\lambda_n},$$

which yields

$$e_n(\mathbf{x}) = f_S(\mathbf{x}).$$

Similarly, h_n is the sum of all monomials of degree n and is the weight generating function for all partitions with n parts. What if we want to count partitions with any number of parts?

Proposition 4.3.5 *We have the following generating functions*

$$E(t) \stackrel{\text{def}}{=} \sum_{n \geq 0} e_n(\mathbf{x}) t^n = \prod_{i \geq 1} (1 + x_i t),$$

$$H(t) \stackrel{\text{def}}{=} \sum_{n \geq 0} h_n(\mathbf{x}) t^n = \prod_{i \geq 1} \frac{1}{(1 - x_i t)}.$$

Proof. Work in the ring $\mathbb{C}[[\mathbf{x}, t]]$. For the elementary symmetric functions, consider the set $S = \{\lambda \ : \ \lambda \text{ with distinct parts}\}$ with weight

$$\text{wt}' \lambda = t^{l(\lambda)} \text{ wt } \lambda,$$

where wt is as before. Then

$$
\begin{aligned}
f_S(\mathbf{x}, t) &= \sum_{\lambda \in S} \text{wt}' \lambda \\
&= \sum_{n \geq 0} \sum_{l(\lambda) = n} t^n \text{ wt } \lambda \\
&= \sum_{n \geq 0} e_n(\mathbf{x}) t^n.
\end{aligned}
$$

To obtain the product, write

$$S = (\{1^0\} \uplus \{1^1\}) \times (\{2^0\} \uplus \{2^1\}) \times (\{3^0\} \uplus \{3^1\}) \times \cdots,$$

so that

$$f_S(\mathbf{x}, t) = (1 + x_1 t)(1 + x_2 t)(1 + x_3 t) \cdots.$$

The proof for the complete symmetric functions is analogous. ∎

While we are computing generating functions, we might as well give one for the power sums. Actually, it is easier to produce one for $p_n(\mathbf{x})/n$.

Proposition 4.3.6 *We have the following generating function:*

$$\sum_{n \geq 1} p_n(\mathbf{x}) \frac{t^n}{n} = \ln \prod_{i \geq 1} \frac{1}{(1 - x_i t)}.$$

Proof. Using the Taylor expansion of $\ln \frac{1}{1-x}$, we obtain

$$
\begin{aligned}
\ln \prod_{i \geq 1} \frac{1}{(1 - x_i t)} &= \sum_{i \geq 1} \ln \frac{1}{(1 - x_i t)} \\
&= \sum_{i \geq 1} \sum_{n \geq 1} \frac{(x_i t)^n}{n} \\
&= \sum_{n \geq 1} \frac{t^n}{n} \sum_{i \geq 1} x_i^n \\
&= \sum_{n \geq 1} p_n(\mathbf{x}) \frac{t^n}{n}. \quad\blacksquare
\end{aligned}
$$

In order to have enough elements for a basis of Λ^n, we must have one function for each partition of n according to Proposition 4.3.3. To extend Definition 4.3.4 to $\lambda = (\lambda_1, \lambda_2, \ldots, \lambda_l)$, let

$$f_\lambda = f_{\lambda_1} f_{\lambda_2} \cdots f_{\lambda_l},$$

where $f = p$, e, or h. We say that these functions are *multiplicative*. To illustrate, if $\lambda = (2, 1)$, then

$$p_{(2,1)} = (x_1^2 + x_2^2 + x_3^2 + \cdots)(x_1 + x_2 + x_3 + \cdots).$$

Theorem 4.3.7 *The following are bases for* Λ^n.

1. $\{p_\lambda \ : \ \lambda \vdash n\}$.

2. $\{e_\lambda \ : \ \lambda \vdash n\}$.

3. $\{h_\lambda \ : \ \lambda \vdash n\}$.

Proof.

1. Let $C = (c_{\lambda\mu})$ be the matrix expressing the p_λ in terms of the basis m_μ. If we can find an ordering of partitions such that C is triangular with nonzero entries down the diagonal, then C^{-1} exists, and the p_λ are also a basis. It turns out that lexicographic order will work. In fact, we claim that

$$p_\lambda = c_{\lambda\lambda} m_\lambda + \sum_{\mu \rhd \lambda} c_{\lambda\mu} m_\mu, \qquad (4.6)$$

where $c_{\lambda\lambda} \neq 0$. (This is actually stronger than our claim about C by Proposition 2.2.6.) But if $\mathbf{x}_1^{\mu_1} \mathbf{x}_2^{\mu_2} \cdots \mathbf{x}_m^{\mu_m}$ appears in

$$p_\lambda = (x_1^{\lambda_1} + x_2^{\lambda_1} + \cdots)(x_1^{\lambda_2} + x_2^{\lambda_2} + \cdots) \cdots$$

then each μ_i must be a sum of λ_j's. Since adding together parts of a partition makes it become larger in dominance order, m_λ must be the smallest term that occurs.

2. In a similar manner we can show that there exist scalars $d_{\lambda\mu}$ such that

$$e_{\lambda'} = m_\lambda + \sum_{\mu \lhd \lambda} d_{\lambda\mu} m_\mu,$$

where λ' is the conjugate of λ.

3. Since there are $p(n) = \dim \Lambda^n$ functions h_λ, it suffices to show that they generate the basis e_μ. Since both sets of functions are multiplicative, we may simply demonstrate that every e_n is a polynomial in the h_k. From the products in Proposition 4.3.5, we see that

$$H(t)E(-t) = 1.$$

Substituting in the summations for H and E and picking out the coefficient of t^n on both sides yields

$$\sum_{r=0}^{n}(-1)^r h_{n-r} e_r = 0$$

for $n \geq 1$. So

$$e_n = h_1 e_{n-1} - h_2 e_{n-2} + \cdots,$$

which is a polynomial in the h's by induction on n. ∎

Part 2 of this theorem is often called the fundamental theorem of symmetric functions and stated as follows: Every symmetric function is a polynomial in the elementary functions e_n.

4.4 Schur Functions

There is a fifth basis for Λ^n that is very important, the Schur functions. As we will see, they are also intimately connected with the irreducible representations of S_n and tableaux. In fact, they are so protean that there are many different ways to define them. In this section, we take the combinatorial approach.

Given any composition $\mu = (\mu_1, \mu_2, \ldots, \mu_l)$, there is a corresponding monomial weight in $\mathbb{C}[[\mathbf{x}]]$:

$$\mathbf{x}^\mu \stackrel{\text{def}}{=} x_1^{\mu_1} x_2^{\mu_2} \cdots x_m^{\mu_l}. \tag{4.7}$$

Now consider any generalized tableau T of shape λ. It also has a weight, namely,

$$\mathbf{x}^T \stackrel{\text{def}}{=} \prod_{(i,j)\in\lambda} x_{T_{i,j}} = \mathbf{x}^\mu, \tag{4.8}$$

where μ is the content of T. For example, if

$$T = \begin{array}{ccc} 4 & 1 & 4 \\ 1 & 3 & \end{array},$$

then

$$\mathbf{x}^T = x_1^2 x_3 x_4^2.$$

Definition 4.4.1 Given a partition λ, the associated *Schur function* is

$$s_\lambda(\mathbf{x}) = \sum_T \mathbf{x}^T,$$

where the sum is over all semistandard λ-tableaux T.

By way of illustration, if $\lambda = (2,1)$, then some of the possible tableaux are

$$T: \begin{array}{cc} 1\,1 \\ 2 \end{array}, \begin{array}{cc} 1\,2 \\ 2 \end{array}, \begin{array}{cc} 1\,1 \\ 3 \end{array}, \begin{array}{cc} 1\,3 \\ 3 \end{array}, \cdots \begin{array}{cc} 1\,2 \\ 3 \end{array}, \begin{array}{cc} 1\,3 \\ 2 \end{array}, \begin{array}{cc} 1\,2 \\ 4 \end{array}, \begin{array}{cc} 1\,4 \\ 2 \end{array}, \cdots,$$

so

$$s_{(2,1)}(\mathbf{x}) = x_1^2 x_2 + x_1 x_2^2 + x_1^2 x_3 + x_1 x_3^2 + \cdots + 2x_1 x_2 x_3 + 2x_1 x_2 x_4 + \cdots.$$

Note that if $\lambda = (n)$, then a one-rowed tableau is just a weakly increasing sequence of n positive integers, i.e., a partition with n parts (written backward), so

$$s_{(n)}(\mathbf{x}) = h_n(\mathbf{x}). \tag{4.9}$$

If we have only one column, then the entries must increase from top to bottom, so the partition must have distinct parts and thus

$$s_{(1^n)} = e_n(\mathbf{x}). \tag{4.10}$$

Finally, if $\lambda \vdash n$ is arbitrary, then

$$[x_1 x_2 \cdots x_n] s_\lambda(\mathbf{x}) = f^\lambda,$$

since pulling out this coefficient merely considers the standard tableaux.

Before we can show that the s_λ are a basis for Λ^n, we must verify that they are indeed symmetric functions. We give two proofs of this fact, one based on our results from representation theory and one combinatorial (the latter being due to Knuth [Knu 70]).

Proposition 4.4.2 *The function $s_\lambda(\mathbf{x})$ is symmetric.*

Proof 1. By definition of the Schur functions and Kostka numbers,

$$s_\lambda = \sum_\mu K_{\lambda\mu} \mathbf{x}^\mu, \tag{4.11}$$

where the sum is over all compositions μ of n. Thus it is enough to show that

$$K_{\lambda\mu} = K_{\lambda\tilde{\mu}} \tag{4.12}$$

for any rearrangement $\tilde{\mu}$ of μ. But in this case M^μ and $M^{\tilde{\mu}}$ are isomorphic modules. Thus they have the same decomposition into irreducibles, and (4.12) follows from Young's rule (Theorem 2.11.2).

Proof 2. It suffices to show that

$$(i, i+1)s_\lambda(\mathbf{x}) = s_\lambda(\mathbf{x})$$

for each adjacent transposition. To this end, we describe an involution on semistandard λ-tableaux

$$T \longrightarrow T'$$

such that the numbers of i's and $(i + 1)$'s are exchanged when passing from T to T' (with all other multiplicities staying the same).

Given T, each column contains either an $i, i+1$ pair; exactly one of $i, i+1$; or neither. Call the pairs *fixed* and all other occurrences of i or $i + 1$ *free*. In each row switch the number of free i's and $(i + 1)$'s; i.e., if the the row consists of k free i's followed by l free $(i + 1)$'s then replace them by l free i's followed by k free $(i + 1)$'s. To illustrate, if $i = 2$ and

$$T = \begin{matrix} 1 & 1 & 1 & 1 & 2 & 2 & 2 & 2 & 2 & 3 \\ 2 & 2 & 3 & 3 & 3 & 3 & & & & \\ 3 & & & & & & & & & \end{matrix} \quad ,$$

then the twos and threes in columns 2 through 4 and 7 through 10 are free. So

$$T' = \begin{matrix} 1 & 1 & 1 & 1 & 2 & 2 & 2 & 3 & 3 & 3 \\ 2 & 2 & 2 & 3 & 3 & 3 & & & & \\ 3 & & & & & & & & & \end{matrix} \quad .$$

The new tableau T' is still semistandard by the definition of free. Since the fixed i's and $(i + 1)$'s come in pairs, this map has the desired exchange property. It is also clearly an involution. ∎

Using the ideas in the proof of Theorem 4.3.7, part 1, the following result guarantees that the s_λ are a basis.

Proposition 4.4.3 *We have*

$$s_\lambda = \sum_{\mu \unlhd \lambda} K_{\lambda\mu} m_\mu,$$

where the sum is over partitions μ (rather than compositions) and $K_{\lambda\lambda} = 1$.

Proof. By equation (4.11) and the symmetry of the Schur functions, we have

$$s_\lambda = \sum_\mu K_{\lambda\mu} m_\mu,$$

where the sum is over all *partitions* μ. We can prove that

$$K_{\lambda\mu} = \begin{cases} 0 & \text{if } \lambda \ntrianglelefteq \mu, \\ 1 & \text{if } \lambda = \mu, \end{cases}$$

in two different ways.

One is to appeal again to Young's rule and Corollary 2.4.7. The other is combinatorial. If $K_{\lambda\mu} \neq 0$, then consider a λ-tableau T of content μ. Since T is column-strict, all occurrences of the numbers $1, 2, \ldots, i$ are in rows 1 through i. This implies that for all i,

$$\mu_1 + \mu_2 + \cdots + \mu_i \le \lambda_1 + \lambda_2 + \cdots + \lambda_i,$$

i.e., $\mu \unlhd \lambda$. Furthermore, if $\lambda = \mu$, then by the same reasoning there is only one tableau of shape and content λ, namely, the one where row i contains all occurrences of i. (Some authors call this tableau *superstandard*.) ∎

Corollary 4.4.4 *The set $\{s_\lambda : \lambda \vdash n\}$ is a basis for Λ^n.* ∎

4.5   The Jacobi-Trudi Determinants

The determinantal formula (Theorem 3.11.1) calculated the number of standard tableaux, f^λ. Analogously, the Jacobi-Trudi determinants provide another expression for s_λ in terms of elementary and complete symmetric functions. Jacobi [Jac 41] was the first to obtain this result, and his student Trudi [Tru 64] subsequently simplified it.

We have already seen the special 1×1 case of these determinants in equations (4.9) and (4.10). The general result is as follows. Any symmetric function with a negative subscript is defined to be zero.

Theorem 4.5.1 (Jacobi-Trudi Determinants) *Let* $\lambda = (\lambda_1, \lambda_2, \ldots, \lambda_l)$. *We have*

$$s_\lambda = |h_{\lambda_i - i + j}|$$

and

$$s_{\lambda'} = |e_{\lambda_i - i + j}|,$$

where λ' *is the conjugate of* λ *and both determinants are* $l \times l$.

Proof. We prove this theorem using a method of Lindström [Lin 73] that was independently discovered and exploited by Gessel [Ges um] and Gessel-Viennot [G-V 85, G-V ip]. (See also Karlin [Kar 88].) The crucial insight is that one can view both tableaux and determinants as lattice paths. Consider the plane $\mathbb{Z} \times \mathbb{Z}$ of integer lattice points. We consider (possibly infinite) paths in this plane

$$p = s_1, s_2, s_3, \ldots,$$

where each step s_i is of unit length northward (N) or eastward (E). Such a path is shown in the following figure.

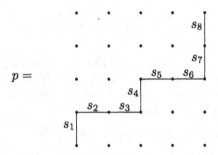

$$p =$$

Label the eastward steps of p using one of two labelings. The *e-labeling* assigns to each eastward s_i the label

$$L(s_i) = i.$$

The *h-labeling* gives s_i the label

$$\check{L}(s_i) = \text{(the number of northward } s_j \text{ preceding } s_i) + 1.$$

Intuitively, in the *h*-labeling all the eastward steps on the line through the origin of p are labeled 1, all those on the line one unit above are labeled 2, and so on. Labeling our example path with each of the two possibilities yields the next pair of diagrams.

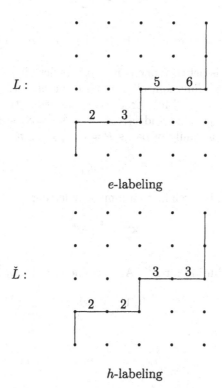

e-labeling

h-labeling

It is convenient to extend $\mathbb{Z} \times \mathbb{Z}$ by the addition of some points at infinity. Specifically, for each $x \in \mathbb{Z}$, add a point (x, ∞) above every point on the vertical line with coordinate x. We assume that a path can reach (x, ∞) only by ending with an infinite number of consecutive northward steps along this line. If p starts at a vertex u and ends at a vertex v (which may be a point at infinity), then we write $u \xrightarrow{p} v$.

There are two weightings of paths corresponding to the two labelings. If p has only a finite number of eastward steps, define

$$\mathbf{x}^p = \prod_{s_i \in p} x_{L(s_i)}$$

and

$$\check{\mathbf{x}}^p = \prod_{s_i \in p} x_{\check{L}(s_i)},$$

where each product is taken over the eastward s_i in p. Note that $\mathbf{x}^p$ is always square-free and $\check{\mathbf{x}}^p$ can be any monomial. So we have

$$e_n(\mathbf{x}) = \sum_p \mathbf{x}^p$$

and

$$h_n(\mathbf{x}) = \sum_p \check{\mathbf{x}}^p,$$

where both sums are over all paths $(a, b) \xrightarrow{p} (a + n, \infty)$ for any fixed initial vertex (a, b).

Just as all paths between one pair of points describes a lone elementary or complete symmetric function, all l-tuples of paths between l pairs of points will be used to model the l-fold products contained in the Jacobi-Trudi determinants. Let $u_1, u_2, \ldots, u_l$ and $v_1, v_2, \ldots, v_l$ be fixed sets of initial and final vertices. Consider a family of paths $\mathcal{P} = (p_1, p_2, \ldots, p_l)$, where, for each i,

$$u_i \xrightarrow{p_i} v_{\pi i}$$

for some $\pi \in S_l$. Give weight to a family by letting

$$\mathbf{x}^{\mathcal{P}} = \prod_i \mathbf{x}^{p_i}$$

with a similar definition for $\check{\mathbf{x}}^{\mathcal{P}}$. Also, define the *sign of* $\mathcal{P}$ to be

$$(-1)^{\mathcal{P}} = \operatorname{sgn} \pi.$$

For example, the family

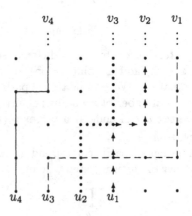

has

$$\check{\mathbf{x}}^{\mathcal{P}} = x_2^4 x_3^2 x_4$$

and sign
$$(-1)^P = \operatorname{sgn}(1,2,3)(4) = +1.$$

We now concentrate on proving
$$s_\lambda = |h_{\lambda_i - i + j}|, \tag{4.13}$$

the details for the elementary determinant being similar and left to the reader. Given λ, pick initial points
$$u_i = (1 - i, 0) \tag{4.14}$$

and the final ones
$$v_i = (\lambda_i - i + 1, \infty). \tag{4.15}$$

The preceding 4-tuple is an example for $\lambda = (2, 2, 2, 1)$. From the choice of vertices,
$$h_{\lambda_i - i + j} = \sum_{u_j \xrightarrow{P} v_i} \check{\mathbf{x}}^P.$$

Thus the set of l-tuples $\mathcal{P}$ with permutation π corresponds to the term in the determinant obtained from the entries with coordinates
$$(\pi 1, 1), \ (\pi 2, 2), \ \ldots, \ (\pi l, l).$$

So
$$|h_{\lambda_i - i + j}(\mathbf{x})| = \sum_{\mathcal{P}} (-1)^P \check{\mathbf{x}}^P, \tag{4.16}$$

where the sum is over all families of paths with initial points and final points given by the u_j and v_i.

Next we show that all the terms on the right side of equation (4.16) cancel in pairs except for those corresponding to l-tuples of nonintersecting paths. To do this, we need a weight-preserving involution
$$\mathcal{P} \xleftrightarrow{\ \iota\ } \mathcal{P}'$$

such that if $\mathcal{P} = \mathcal{P}'$ (corresponding to the 1-cycles of ι), then $\mathcal{P}$ is noninter-secting and if $\mathcal{P} \neq \mathcal{P}'$ (corresponding to the 2-cycles), then $(-1)^P = -(-1)^{P'}$. The basic idea is that if $\mathcal{P}$ contains some intersections, then we will find two uniquely defined intersecting paths and switch their final portions.

Definition 4.5.2 Given $\mathcal{P}$, define $\iota \mathcal{P} = \mathcal{P}'$, where

1. if $p_i \cap p_j = \emptyset$ for all i, j, then $\mathcal{P} = \mathcal{P}'$,

2. otherwise, find the smallest index i such that the path p_i intersects some other path. Let v_0 be the first (SW-most) intersection on p_i and let p_j be the other path through v_0. (If there is more than one, choose p_j such that j is minimal.) Now define
$$\mathcal{P}' = \mathcal{P} \quad \text{with } p_i, p_j \text{ replaced by } p_i', p_j',$$

where
$$p_i' = u_i \xrightarrow{p_i} v_0 \xrightarrow{p_j} v_{\pi j} \quad \text{and} \quad p_j' = u_j \xrightarrow{p_j} v_0 \xrightarrow{p_i} v_{\pi i}. \ \blacksquare$$

By way of illustration, we can apply this map to our 4-tuple:

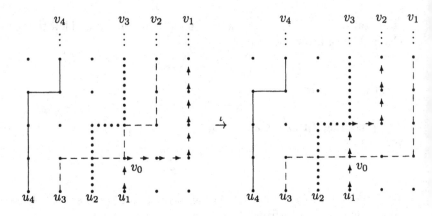

Our choice of p_i, p_j, and v_0 makes it possible to reconstruct them after applying ι, so the map is invertible, with itself as inverse. The nonintersecting families are fixed by definition. For the intersecting l-tuples, the sign clearly changes in passing from $\mathcal{P}$ to $\mathcal{P}'$. Weight is preserved due to the fact that all paths start on the same horizontal axis. Thus the set of labels is unchanged by ι.

Because of the cancellation in (4.16), the only thing that remains is to show that

$$s_\lambda(\mathbf{x}) = \sum_{\mathcal{P}} (-1)^{\mathcal{P}} \check{\mathbf{x}}^{\mathcal{P}},$$

where the sum is now over all l-tuples of nonintersecting paths. But by our choice of initial and final vertices, $\mathcal{P} = (p_1, p_2, \ldots, p_l)$ is nonintersecting only if it corresponds to the identity permutation. Thus $(-1)^{\mathcal{P}} = +1$ and $u_i \xrightarrow{p_i} v_i$ for all i. There is a simple bijection between such families and semistandard tableaux. Given $\mathcal{P}$, merely use the h-labels of the ith path, listed in increasing order, for the ith row of a tableau T. For example,

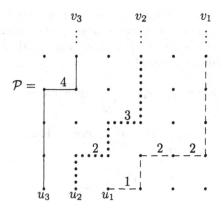

becomes the tableau

$$T = \begin{array}{ccc} 1 & 2 & 2 \\ 2 & 3 & \\ 4 & & \end{array}.$$

In view of equations (4.14) and (4.15), T has shape λ. By definition the rows of T weakly increase. Finally, the columns must strictly increase because the nonintersecting condition and choice of initial vertices force the jth eastward step of p_{i+1} to be higher than the corresponding step on p_i. Construction of the inverse map is an easy exercise. This completes the proof of (4.13). ∎

4.6 Other Definitions of the Schur Function

We would be remiss if we did not give the definition that Schur originally used [Scu 01] (as a quotient of alternants) for the functions that bear his name. This definition actually goes back to Jacobi [Jac 41], but Schur was the first to notice the connection with irreducible characters of $\mathcal{S}_n$, which is our third way of defining s_λ.

First, we have to restrict Λ to l variables, where $l = l(\lambda)$. (Actually any $l \geq l(\lambda)$ will do for s_λ.) Specifically, let

$$\Lambda_l = \{f(x_1, \ldots, x_l, 0, 0, \ldots) \ : \ f(\mathbf{x}) \in \Lambda\}.$$

We use the abbreviation $f(x_1, x_2, \ldots, x_l)$ for a typical element of Λ_l. For example,

$$p_3(x_1, x_2, x_3, x_4) = x_1^3 + x_2^3 + x_3^3 + x_4^3.$$

Thus Λ_l is just the set of functions in the polynomial ring $\mathbb{C}[x_1, x_2, \ldots, x_l]$ that are fixed by the action (4.4) of $\mathcal{S}_l$.

Similarly, we can consider *skew-symmetric functions* in $\mathbb{C}[x_1, x_2, \ldots, x_l]$, which are those satisfying

$$\pi f = (\operatorname{sgn} \pi) f$$

for all $\pi \in \mathcal{S}_l$. Just as we can obtain symmetric functions by symmetrizing a monomial, we can obtain skew-symmetric ones by skew-symmetrization. Let $\mu = (\mu_1, \mu_2, \ldots, \mu_l)$ be any composition with monomial $\mathbf{x}^\mu$ given by (4.7). Define the corresponding *alternant* by

$$a_\mu(x_1, \ldots, x_l) = \sum_{\pi \in \mathcal{S}_l} (\text{sgn } \pi) \pi \mathbf{x}^\mu.$$

It is easy to verify that a_μ is skew-symmetric. From the definition of a determinant, we can also write

$$a_\mu = \left| x_i^{\mu_j} \right|_{1 \leq i, j \leq l}. \tag{4.17}$$

For example,

$$a_{(4,2,1)}(x_1, x_2, x_3) = x_1^4 x_2^2 x_3 + x_1^2 x_2 x_3^4 + x_1 x_2^4 x_3^2 - x_1^4 x_2 x_3^2 - x_1^2 x_2^4 x_3 - x_1 x_2^2 x_3^4$$

$$= \begin{vmatrix} x_1^4 & x_1^2 & x_1 \\ x_2^4 & x_2^2 & x_2 \\ x_3^4 & x_3^2 & x_3 \end{vmatrix}.$$

The most famous example of an alternant arises when we let the composition be

$$\delta \stackrel{\text{def}}{=} (l - 1, l - 2, \ldots, 1, 0).$$

In this case

$$a_\delta = \left| x_i^{l-j} \right|$$

is the *Vandermonde determinant*. It is well-known that we have the factorization

$$a_\delta = \prod_{1 \leq i < j \leq l} (x_i - x_j). \tag{4.18}$$

If λ is any partition of length l with two equal parts, then $a_\lambda = 0$, since determinant (4.17) has two equal columns. So it does no harm to restrict our attention to partitions with distinct parts. These all have the form $\lambda + \delta$, where λ is an arbitrary partition and addition of integer vectors is done componentwise. Furthermore, if we set $x_i = x_j$ in any alternant, then we have two equal rows. So $a_{\lambda+\delta}$ is divisible by all the terms in product (4.18), which makes $a_{\lambda+\delta}/a_\delta$ a polynomial. In fact, it must be symmetric, being the quotient of skew-symmetric polynomials. The amazing thing, at least from our point of view, is that this is actually the symmetric function s_λ. To demonstrate this we follow the proof in Macdonald [Mac 79, page 25]. For $1 \leq j \leq l$, let $e_n^{(j)}$ denote the elementary symmetric function in the variables $x_1, \ldots, x_{j-1}, x_{j+1}, \ldots, x_l$ (x_j omitted). We have the following lemma.

Lemma 4.6.1 *Let* $\mu = (\mu_1, \mu_2, \ldots, \mu_l)$ *be any composition. Consider the* $l \times l$ *matrices*

$$A_\mu = (x_j^{\mu_i}), \quad H_\mu = (h_{\mu_i - l + j}) \quad and \quad E = ((-1)^{l-i} e_{l-i}^{(j)}).$$

Then

$$A_\mu = H_\mu E.$$

Proof. Consider the generating function for the $e_n^{(j)}$,

$$E^{(j)}(t) \stackrel{\text{def}}{=} \sum_{n=0}^{l-1} e_n^{(j)} t^n = \prod_{i \neq j} (1 + x_i t).$$

We can now mimic the proof of Theorem 4.3.7, part 3. Since

$$H(t) E^{(j)}(-t) = \frac{1}{1 - x_j t},$$

we can extract the coefficient of t^{μ_i} on both sides. This yields

$$\sum_{k=1}^{l} h_{\mu_i - l + k} \cdot (-1)^{l-k} e_{l-k}^{(j)} = x_j^{\mu_i},$$

which is equivalent to what we wished to prove. ∎

Corollary 4.6.2 *Let λ have length l. Then*

$$s_\lambda = \frac{a_{\lambda + \delta}}{a_\delta},$$

where all functions are polynomials in $x_1, \ldots, x_l$.

Proof. Taking determinants in the lemma, we obtain

$$|A_\mu| = |H_\mu| \cdot |E|, \tag{4.19}$$

where $|A_\mu| = a_\mu$. First of all, let $\mu = \delta$. In this case $H_\delta = (h_{i-j})$, which is upper unitriangular and thus has determinant 1. Plugging this into (4.19) gives $|E| = a_\delta$.

Now, letting $\mu = \lambda + \delta$ in the same equation, we have

$$\frac{a_{\lambda + \delta}}{a_\delta} = |H_{\lambda + \delta}| = |h_{\lambda_i - i + j}|.$$

Hence we are done, by the Jacobi-Trudi theorem. ∎

Our last description of the s_λ will involve the characters of S_n. To see the connection, let us reexamine the change-of-basis matrix between the monomial and power sum symmetric functions introduced in (4.6). Let us compute a small example. When $n = 3$, we obtain

$$
\begin{array}{rclcl}
p_{(3)} & = & x_1^3 + x_2^3 + \cdots & = & m_{(3)}, \\
p_{(2,1)} & = & (x_1^2 + x_2^2 + \cdots)(x_1 + x_2 + \cdots) & = & m_{(3)} + m_{(2,1)}, \\
p_{(1^3)} & = & (x_1 + x_2 + \cdots)^3 & = & m_{(3)} + 3m_{(2,1)} + 6m_{(1^3)}.
\end{array}
$$

Comparing this with the character values ϕ^μ of the permutation modules M^μ in Example 2.1.9, the reader will be led to suspect the following theorem.

Theorem 4.6.3 *Let ϕ_λ^μ be the character of M^μ evaluated on the class corresponding to λ. Then*

$$p_\lambda = \sum_{\mu \trianglerighteq \lambda} \phi_\lambda^\mu m_\mu. \tag{4.20}$$

Proof. Let $\lambda = (\lambda_1, \lambda_2, \ldots, \lambda_l)$. Then we can write equation (4.6) as

$$\prod_i (x_1^{\lambda_i} + x_2^{\lambda_i} + \cdots) = \sum_\mu c_{\lambda\mu} m_\mu.$$

Pick out the coefficient of $\mathbf{x}^\mu$ on both sides, where $\mu = (\mu_1, \mu_2, \ldots, \mu_m)$. On the right, it is $c_{\lambda\mu}$. On the left, it is the number of ways to distribute the parts of λ into subpartitions $\lambda^1, \ldots, \lambda^m$ such that

$$\biguplus_i \lambda^i = \lambda \quad \text{and} \quad \lambda^i \vdash \mu_i \quad \text{for all } i, \tag{4.21}$$

where equal parts of λ are distinguished in order to be considered different in the disjoint union.

Now consider $\phi_\lambda^\mu = \phi^\mu(\pi)$, where $\pi \in \mathcal{S}_n$ is an element of cycle type λ. By definition, this character value is the number of fixed-points of the action of π on all standard tabloids t of shape μ. But t is fixed if and only if each cycle of π lies in a single row of t. Thus we must distribute the cycles of length λ_i among the rows of length μ_j subject to exactly the same restrictions as in (4.21). It follows that $c_{\lambda\mu} = \phi_\lambda^\mu$, as desired. ∎

Equation (4.20) shows that p_λ is the generating function for the character values on a fixed conjugacy class K_λ as the module in question varies over all M^μ. What we would like is a generating function for the fixed character χ^λ of the irreducible Specht module S^λ as K_μ varies over all conjugacy classes. To isolate χ^λ in (4.20), we use the inner product

$$\langle \phi, \chi \rangle = \frac{1}{n!} \sum_{\pi \in \mathcal{S}_n} \phi(\pi)\chi(\pi) = \text{ the multiplicity of } \chi \text{ in } \phi$$

as long as χ is irreducible (Corollary 1.9.4, part 2).

If $\pi \in \mathcal{S}_n$ has type λ, then it will be convenient to define the corresponding power sum symmetric function $p_\pi = p_\lambda$; similar definitions can be made for the other bases. For example, if $\pi = (1, 3, 4)(2, 5)$, then $p_\pi = p_{(3,2)}$. From the previous theorem,

$$p_\pi = \sum_\mu \phi^\mu(\pi) m_\mu.$$

To introduce the inner product, we multiply by $\chi^\lambda(\pi)/n!$ and sum to get

$$
\begin{aligned}
\frac{1}{n!} \sum_{\pi \in \mathcal{S}_n} p_\pi \chi^\lambda(\pi) &= \frac{1}{n!} \sum_{\pi \in \mathcal{S}_n} \left(\sum_\mu \phi^\mu(\pi) m_\mu \right) \chi^\lambda(\pi) \\
&= \sum_\mu m_\mu \left(\frac{1}{n!} \sum_\pi \phi^\mu(\pi) \chi^\lambda(\pi) \right) \\
&= \sum_\mu K_{\lambda\mu} m_\mu \qquad \text{(Young's rule)} \\
&= s_\lambda. \qquad\qquad\qquad \text{(Proposition 4.4.3)}
\end{aligned}
$$

We have proved the following theorem of Frobenius.

Theorem 4.6.4 *If* $\lambda \vdash n$, *then*

$$s_\lambda = \frac{1}{n!} \sum_{\pi \in \mathcal{S}_n} \chi^\lambda(\pi) p_\pi. \quad \blacksquare \tag{4.22}$$

Note that this is a slightly different type of generating function from those discussed previously, using the power sum basis and averaging over $\mathcal{S}_n$. Such functions occur in other areas of combinatorics, notably Pólya theory, where they are related to cycle index polynomials of groups [PTW 83, pages 55–85].

There are several other ways to write equation (4.22). Since χ^λ is a class function, we can collect terms and obtain

$$s_\lambda = \frac{1}{n!} \sum_\mu k_\mu \chi_\mu^\lambda p_\mu,$$

where $k_\mu = |K_\mu|$ and χ_μ^λ is the value of χ^λ on K_μ. Alternatively, we can use formula (1.2) to express s_λ in terms of the size of centralizers:

$$s_\lambda = \sum_\mu \frac{1}{z_\mu} \chi_\mu^\lambda p_\mu. \tag{4.23}$$

4.7 The Characteristic Map

Let $R^n = R(\mathcal{S}_n)$ be the space of class functions on $\mathcal{S}_n$. Then there is an intimate connection between R^n and Λ^n that we will explore in this section.

First of all, $\dim R^n = \dim \Lambda^n = p(n)$ (the number of partitions of n), so these two are isomorphic as vector spaces. We also have an inner product on R^n for which the irreducible characters on $\mathcal{S}_n$ form an orthonormal basis (Theorem 1.9.3). Motivated by equation (4.22), define an inner product on Λ^n by

$$\langle s_\lambda, s_\mu \rangle = \delta_{\lambda\mu}$$

and sesquilinear extension (linear in the first variable and conjugate linear in the second).

We now define a map to preserve these inner products.

Definition 4.7.1 The *characteristic map* is $\mathrm{ch}^n : R^n \to \Lambda^n$ defined by

$$\mathrm{ch}^n(\chi) = \sum_{\mu \vdash n} z_\mu^{-1} \chi_\mu p_\mu,$$

where χ_μ is the value of χ on the class μ.

It is easy to verify that ch^n is linear. Furthermore, if we apply ch^n to the irreducible characters, then by equation (4.23)

$$\mathrm{ch}^n(\chi^\lambda) = s_\lambda.$$

Since ch^n takes one orthonormal basis to another, we immediately have the following.

Proposition 4.7.2 *The map* ch^n *is an isometry between* R^n *and* Λ^n. ∎

Now consider $R = \oplus_n R^n$, which is isomorphic to $\Lambda = \oplus_n \Lambda^n$ via the characteristic map $\mathrm{ch} = \oplus_n \mathrm{ch}^n$. But Λ also has the structure of a graded algebra—i.e., a ring product satisfying (4.5). How can we construct a corresponding product in R^n? If χ and ψ are characters of $\mathcal{S}_n$ and $\mathcal{S}_m$, respectively, we want to produce a character of $\mathcal{S}_{n+m}$. But the tensor product $\chi \otimes \psi$ gives us a character of $\mathcal{S}_n \times \mathcal{S}_m$, and induction gets us into the group we want. Therefore, define a product on R by bilinearly extending

$$\chi \cdot \psi = (\chi \otimes \psi){\uparrow}^{\mathcal{S}_{n+m}},$$

where χ and ψ are characters.

Before proving that this product agrees with the one in Λ, we must generalize some of the concepts we have previously introduced. Let G be any group and let $\mathcal{A}$ be any algebra over $\mathbb{C}$. Consider functions $\chi, \psi : G \to \mathcal{A}$ with the *bilinear form*

$$\langle \chi, \psi \rangle' = \frac{1}{|G|} \sum_{g \in G} \chi(g)\psi(g^{-1}).$$

Note that since g and g^{-1} are in the same conjugacy class in $\mathcal{S}_n$, we have, for any class function χ,

$$\mathrm{ch}^n(\chi) = \frac{1}{n!} \sum_{\pi \in \mathcal{S}_n} \chi(\pi)p_\pi = \langle \chi, p \rangle',$$

where $p : \mathcal{S}_n \to \Lambda^n$ is the function $p(\pi) = p_\pi$.

Suppose $H \leq G$. If $\chi : G \to \mathcal{A}$, then define the *restriction of* χ *to* H to be the map $\chi{\downarrow}_H : H \to \mathcal{A}$ such that

$$\chi{\downarrow}_H (h) = \chi(h)$$

for all $h \in H$. On the other hand, if $\psi : H \to \mathcal{A}$, then let the *induction of* ψ *to* G be $\psi{\uparrow}^G : G \to \mathcal{A}$ defined by

$$\psi{\uparrow}^G (g) = \frac{1}{|H|} \sum_{x \in G} \psi(x^{-1}gx),$$

where $\psi = 0$ outside of H. The reader should verify that the following generalization of Frobenius reciprocity (Theorem 1.12.6) holds in this setting.

Theorem 4.7.3 *Consider* $H \leq G$ *with functions* $\chi : G \to \mathcal{A}$ *and* $\psi : H \to \mathcal{A}$. *If* χ *is a class function on* G, *then*

$$\langle \psi{\uparrow}^G, \chi \rangle' = \langle \psi, \chi{\downarrow}_H \rangle'. \ ∎$$

This is the tool we need to get at our main theorem about the characteristic map.

Theorem 4.7.4 *The map* ch $: R \to \Lambda$ *is an isomorphism of algebras.*

Proof. By Proposition 4.7.2, it suffices to check that products are preserved. If χ and ψ are characters in $\mathcal{S}_n$ and $\mathcal{S}_m$, respectively, then using part 2 of Theorem 1.11.2 yields

$$
\begin{aligned}
\mathrm{ch}(\chi \cdot \psi) &= \langle \chi \cdot \psi, p \rangle' \\
&= \langle (\chi \otimes \psi) \!\uparrow^{\mathcal{S}_{n+m}}, p \rangle' \\
&= \langle \chi \otimes \psi, p \!\downarrow_{\mathcal{S}_n \times \mathcal{S}_m} \rangle' \\
&= \frac{1}{n!m!} \sum_{\pi\sigma \in \mathcal{S}_n \times \mathcal{S}_m} (\chi \otimes \psi)(\pi\sigma) p_{\pi\sigma} \\
&= \frac{1}{n!m!} \sum_{\substack{\pi \in \mathcal{S}_n \\ \sigma \in \mathcal{S}_m}} \chi(\pi)\psi(\sigma) p_\pi p_\sigma \\
&= \left[\frac{1}{n!} \sum_{\pi \in \mathcal{S}_n} \chi(\pi) p_\pi \right] \left[\frac{1}{m!} \sum_{\sigma \in \mathcal{S}_m} \psi(\sigma) p_\sigma \right] \\
&= \mathrm{ch}(\chi)\,\mathrm{ch}(\psi). \quad\blacksquare
\end{aligned}
$$

4.8 Knuth's Algorithm

Schensted [Sch 61] realized that his algorithm could be generalized to the case where the first output tableau is semistandard. Knuth [Knu 70] took this one step further, showing how to get a Robinson-Schensted map when both tableaux allow repetitions and thus making connection with the Cauchy identity [Lit 50, p. 103]. Many of the properties of the original algorithm are preserved. In addition, one obtains a new procedure that is dual to the first.

Just as we are to allow repetitions in our tableaux, we must also permit them in our permutations.

Definition 4.8.1 A *generalized permutation* is a two-line array of positive integers

$$
\pi = \begin{matrix} i_1 & i_2 & \cdots & i_n \\ j_1 & j_2 & \cdots & j_n \end{matrix}
$$

whose columns are in lexicographic order, with the top entry taking precedence. Let $\hat{\pi}$ and $\check{\pi}$ stand for the top and bottom rows of π, respectively. The set of all generalized permutations is denoted by GP. $\blacksquare$

Note that the lexicographic condition can be restated: $\hat{\pi}$ is weakly increasing and in $\check{\pi}$, $j_k \le j_{k+1}$ whenever $i_k = i_{k+1}$. An example of a generalized permutation is

$$
\pi = \begin{matrix} 1 & 1 & 1 & 2 & 2 & 3 \\ 2 & 3 & 3 & 1 & 2 & 1 \end{matrix}. \tag{4.24}
$$

If T is any tableau, then let cont T be the content of T. If $\pi \in$ GP, then $\hat{\pi}$ and $\check{\pi}$ can be viewed as (one-rowed) tableaux, so their contents are defined. In the preceding example cont $\hat{\pi} = (3,2,1)$ and cont $\check{\pi} = (2,2,2)$.

The Robinson-Schensted-Knuth correspondence is as follows.

Theorem 4.8.2 ([Knu 70]) *There is a bijection between generalized permutations and pairs of semistandard tableaux of the same shape,*

$$\pi \overset{R-S-K}{\longleftrightarrow} (T, U)$$

such that cont $\tilde{\pi} =$ cont T *and* cont $\hat{\pi} =$ cont U.

Proof. "$\pi \overset{R-S-K}{\longrightarrow} (T, U)$" We form, as before, a sequence of tableau pairs

$$(T_0, U_0) = (\phi, \phi), \ (T_1, U_1), \ (T_2, U_2), \ \ldots, \ (T_n, U_n) = (T, U),$$

where the elements of $\tilde{\pi}$ are inserted into the T's and the elements of $\hat{\pi}$ are placed in the U's. The rules of insertion and placement are exactly the same as for tableaux without repetitions. Applying this algorithm to the preceding permutation, we obtain

$$
\begin{array}{llllllll}
T_i: & \phi, & 2, & 2\,3, & 2\,3\,3, & 1\,3\,3, & 1\,2\,3, & 1\,1\,3 \\
 & & & & 2 & & 2\,3 & 2\,2 & = T, \\
 & & & & & & & 3 &
\end{array}
$$

$$
\begin{array}{llllllll}
U_i: & \phi, & 1, & 1\,1, & 1\,1\,1, & 1\,1\,1, & 1\,1\,1, & 1\,1\,1 \\
 & & & & 2 & & 2\,2 & 2\,2 & = U. \\
 & & & & & & & 3 &
\end{array}
$$

It is easy to verify that the insertion rules ensure semistandardness of T. Also, U has weakly increasing rows because $\hat{\pi}$ is weakly increasing. To show that U's columns strictly increase, we must make sure that no two equal elements of $\hat{\pi}$ can end up in the same column. But if $i_k = i_{k+1} = i$ in the upper row, then we must have $j_k \leq j_{k+1}$. This implies that the insertion path for j_{k+1} will always lie strictly to the right of the path for j_k, which implies the desired result. Note that we have shown that all elements equal to i are placed in U from left to right as the algorithm proceeds.

"$(T, U) \overset{K-S-R}{\longrightarrow} \pi$" Proceed as in the standard case. One problem is deciding which of the maximum elements of U corresponds to the last insertion. But from the observation just made, the rightmost of these maxima is the correct choice with which to start the deletion process.

We also need to verify that the elements removed from T corresponding to equal elements in U come out in weakly decreasing order. This is an easy exercise left to the reader. ∎

Knuth's original formulation of this algorithm had a slightly different point of departure. Just as one can consider a permutation as a matrix of zeros and ones, generalized permutations can be viewed as matrices with nonnegative integral entries. Specifically, with each $\pi \in$ GP, we can associate a matrix of $M = M(\pi)$ with (i, j) entry

$$M_{i,j} = \text{the number of times } \binom{i}{j} \text{ occurs as a column of } \pi.$$

Our example permutation has

$$M(\pi) = \begin{pmatrix} 0 & 1 & 2 \\ 1 & 1 & 0 \\ 1 & 0 & 0 \end{pmatrix}.$$

We can clearly reverse this process, so we have a bijection

$$\pi \longleftrightarrow M$$

between GP and Mat = Mat(**N**), the set of all matrices with nonnegative integral entries and no final rows or columns of zeros (which contribute nothing to π). Note that in translating from π to M, the number of i's in $\hat{\pi}$ (respectively, the number of j's in $\check{\pi}$) becomes the sum of row i (respectively, column j) of M. Thus Theorem 4.8.2 can be restated.

Corollary 4.8.3 *There is a bijection between matrices $M \in$ Mat and pairs of semistandard tableaux of the same shape,*

$$M \overset{\text{R-S-K}}{\longleftrightarrow} (T, U),$$

such that cont T *and* cont U *give the vectors of column and row sums, respectively, of* M. ∎

To translate these results into generating functions, we use two sets of variables, $\mathbf{x} = \{x_1, x_2, \ldots\}$ and $\mathbf{y} = \{y_1, y_2, \ldots\}$. Let the weight of a generalized permutation π be

$$\text{wt } \pi = \mathbf{x}^{\hat{\pi}} \mathbf{y}^{\check{\pi}} = \mathbf{x}^{\text{cont } \hat{\pi}} \mathbf{y}^{\text{cont } \check{\pi}}.$$

To illustrate, the permutation in (4.24) has

$$\text{wt } \pi = x_1^3 x_2^2 x_3 y_1^2 y_2^2 y_3^2.$$

If the column $\binom{i}{j}$ occurs k times in π, then it gives a contribution of $x_i^k y_j^k$ to the weight. Thus the generating function for generalized permutations is

$$\sum_{\pi \in \text{GP}} \text{wt } \pi = \prod_{i,j \geq 1} \sum_{k \geq 0} x_i^k y_j^k = \prod_{i,j \geq 1} \frac{1}{1 - x_i y_j}.$$

As for tableau pairs (T, U), let

$$\text{wt}(T, U) = \mathbf{x}^U \mathbf{y}^T = \mathbf{x}^{\text{cont } U} \mathbf{y}^{\text{cont } T}$$

as in equation (4.8). Restricting to pairs of the same shape,

$$\sum_{\text{sh } T = \text{sh } U} \text{wt}(T, U) = \sum_{\lambda} \left(\sum_{\text{sh } U = \lambda} \mathbf{x}^U \right) \left(\sum_{\text{sh } T = \lambda} \mathbf{y}^T \right) = \sum_{\lambda} s_\lambda(\mathbf{x}) s_\lambda(\mathbf{y}).$$

Since the Robinson-Schensted-Knuth map is weight preserving, we have given a proof of Cauchy's formula.

Theorem 4.8.4 ([Lit 50]) *We have*

$$\sum_\lambda s_\lambda(\mathbf{x})s_\lambda(\mathbf{y}) = \prod_{i,j\geq 1} \frac{1}{1 - x_iy_j}. \quad \blacksquare$$

Note that just as the Robinson-Schensted correspondence is gotten by restricting Knuth's generalization to the case where all entries are distinct, we can obtain

$$n! = \sum_{\lambda\vdash n}(f^\lambda)^2$$

by taking the coefficient of $x_1 \cdots x_n y_1 \cdots y_n$ on both sides of Theorem 4.8.4.

Because the semistandard condition does not treat rows and columns uniformly, there is a second algorithm related to the one just given. It is called the *dual map*. (This is a different notion of "dual" from the one introduced in Chapter 3, e.g., Definition 3.6.8.) Let GP$'$ denote all those permutations in GP where no column is repeated. These correspond to the 0–1 matrices in Mat.

Theorem 4.8.5 ([Knu 70]) *There is a bijection between $\pi \in$ GP$'$ and pairs (T,U) of tableaux of the same shape with T,U^t semistandard,*

$$\pi \overset{R-S-K'}{\longleftrightarrow} (T,U),$$

such that cont $\check{\pi} =$ cont T *and* cont $\hat{\pi} =$ cont U.

Proof. "$\pi \overset{R-S-K'}{\longrightarrow} (T,U)$" We merely replace row insertion in the R-S-K correspondence with a modification of column insertion. This is done by insisting that at each stage the element entering a column displaces the smallest entry greater than *or equal* to it. For example,

$$c_2 \begin{pmatrix} 1 & 1 & 3 \\ 2 & 3 \\ 3 \end{pmatrix} = \begin{matrix} 1 & 1 & 3 & 3 \\ 2 & 2 \\ 3 \end{matrix} \quad .$$

Note that this is exactly what is needed to ensure that T will be column-strict. The fact that U will be row-strict follows because a subsequence of $\check{\pi}$ corresponding to equal elements in $\hat{\pi}$ must be strictly increasing (since $\pi \in$ GP$'$).

"$(T,U) \overset{K-S-R'}{\longrightarrow} \pi$" The details of the step-by-step reversal and verification that $\pi \in$ GP$'$ are routine. $\blacksquare$

Taking generating functions with the same weights as before yields the dual Cauchy identity.

Theorem 4.8.6 ([Lit 50]) *We have*

$$\sum_\lambda s_\lambda(\mathbf{x})s_{\lambda'}(\mathbf{y}) = \prod_{i,j\geq 1}(1 + x_iy_j),$$

where λ' is the conjugate of λ. $\blacksquare$

Most of the results of Chapter 3 about Robinson-Schensted have generalizations for the Knuth map. We survey a few of them next.

Taking the inverse of a permutation corresponds to transposing the associated permutation matrix. So the following strengthening of Schützenberger's Theorem 3.6.6 should come as no surprise.

Theorem 4.8.7 *If $M \in$ Mat and $M \overset{R-S-K}{\longleftrightarrow} (T, U)$, then*

$$M^t \overset{R-S-K}{\longleftrightarrow} (U, T). \blacksquare$$

We can also deal with the reversal of a generalized permutation. Row and modified column insertion commute as in Proposition 3.2.2. So we obtain the following analogue of Theorem 3.2.3.

Theorem 4.8.8 *If $\pi \in$ GP, then $T(\tilde{\pi}^r) = T'(\tilde{\pi})$, where T' denotes modified column insertion.* $\blacksquare$

The Knuth relations become

$$\text{replace } xzy \text{ by } zxy \text{ if } x \le y < z$$

and

$$\text{replace } yxz \text{ by } yzx \text{ if } x < y \le z.$$

Theorem 3.4.3 remains true.

Theorem 4.8.9 ([Knu 70]) *A pair of generalized permutations are Knuth equivalent if and only if they have the same T-tableau.* $\blacksquare$

Putting together the last two results, we can prove a stronger version of Greene's theorem.

Theorem 4.8.10 ([Gre 74]) *Given $\pi \in$ GP, let $\operatorname{sh} T(\pi) = (\lambda_1, \lambda_2, \ldots, \lambda_l)$ with conjugate $(\lambda_1', \lambda_2', \ldots, \lambda_m')$. Then for any k, $\lambda_1 + \lambda_2 + \cdots + \lambda_k$ and $\lambda_1' + \lambda_2' + \cdots + \lambda_k'$ give the lengths of the longest weakly k-increasing and strictly k-decreasing subsequences of π, respectively.* $\blacksquare$

For the jeu de taquin, we need to break ties when the two elements of T adjacent to the cell to be filled are equal. The correct choice is forced on us by semistandardness. In this case, both the forward and backward slides always move the element that changes rows rather than the one that would change columns. The fundamental results of Schützenberger continue to hold (see Theorems 3.7.7 and 3.7.8).

Theorem 4.8.11 ([Scü 76]) *Let T and U be skew semistandard tableaux. Then T and U have Knuth equivalent row words if and only if they are connected by a sequence of slides. Furthermore, any such sequence bringing them to normal shape results in the first output tableau of the Robinson-Schensted-Knuth correspondence.* $\blacksquare$

Finally, we can define dual equivalence, $\stackrel{*}{\cong}$, in exactly the same way as before (Definition 3.8.2). The result concerning this relation, analogous to Proposition 3.8.1 and Theorem 3.8.8, needed for the Littlewood-Richardson rule is the following.

Theorem 4.8.12 *If T and U are semistandard of the same normal shape, then $T \stackrel{*}{\cong} U$.* ∎

4.9   The Littlewood-Richardson Rule

The Littlewood-Richardson rule gives a combinatorial interpretation to the coefficients of the product $s_\mu s_\nu$ when expanded in terms of the Schur basis. This can be viewed as a generalization of Young's rule, as follows.

We know (Theorem 2.11.2) that

$$M^\mu \cong \bigoplus_\lambda K_{\lambda\mu} S^\lambda, \qquad (4.25)$$

where $K_{\lambda\mu}$ is the number of semistandard tableaux of shape λ and content μ. We can look at this formula from two other perspectives: in terms of characters or symmetric functions.

If $\mu \vdash n$, then M^μ is a module for the induced character $1_{S_\mu} \uparrow^{S_n}$. But from the definitions of the trivial character and the tensor product, we have

$$1_{S_\mu} = 1_{S_{\mu_1}} \otimes 1_{S_{\mu_2}} \otimes \cdots \otimes 1_{S_{\mu_m}},$$

where $\mu = (\mu_1, \mu_2, \ldots, \mu_m)$. Using the product in the class function algebra R (and the transitivity of induction, Exercise 18 of Chapter 1), we can rewrite (4.25) as

$$1_{S_{\mu_1}} \cdot 1_{S_{\mu_2}} \cdots 1_{S_{\mu_m}} = \sum_\lambda K_{\lambda\mu} \chi^\lambda.$$

To bring in symmetric functions, apply the characteristic map to the previous equation (remember that the trivial representation corresponds to an irreducible whose diagram has only one row):

$$s_{(\mu_1)} s_{(\mu_2)} \cdots s_{(\mu_m)} = \sum_\lambda K_{\lambda\mu} s_\lambda.$$

For example,

$$M^{(3,2)} = S^{(3,2)} + S^{(4,1)} + S^{(5)}$$

with the relevant tableaux being

$$T : \quad \begin{array}{ccc} 1\,1\,1 \\ 2\,2 \end{array} \,, \quad \begin{array}{ccc} 1\,1\,1\,2 \\ 2 \end{array} \,, \quad 1\,1\,1\,2\,2 \ .$$

This can be rewritten as

$$1_{S_3} \cdot 1_{S_2} = \chi^{(3,2)} + \chi^{(4,1)} + \chi^{(5)}$$

or

$$s_{(3)}s_{(2)} = s_{(3,2)} + s_{(4,1)} + s_{(5)}.$$

What happens if we try to compute the expansion

$$s_\mu s_\nu = \sum_\lambda c_{\mu\nu}^\lambda s_\lambda, \tag{4.26}$$

where μ and ν are arbitrary partitions? Equivalently, we are asking for the multiplicities of the irreducibles in

$$\chi^\mu \cdot \chi^\nu = \sum_\lambda c_{\mu\nu}^\lambda \chi^\lambda$$

or

$$(S^\mu \otimes S^\nu)\uparrow^{S_n} = \bigoplus_\lambda c_{\mu\nu}^\lambda S^\lambda,$$

where $|\mu| + |\nu| = n$. The $c_{\mu\nu}^\lambda$ are called the *Littlewood-Richardson coefficients.* The importance of the Littlewood-Richardson rule that follows is that it gives a way to interpret these coefficients combinatorially, just as Young's rule does for one-rowed partitions.

We need to explore one other place where these coefficients arise: in the expansion of *skew Schur functions.* Obviously, the definition of $s_\lambda(\mathbf{x})$ given in Section 4.4 makes sense if λ is replaced by a skew diagram. Furthermore, the resulting function $s_{\lambda/\mu}(\mathbf{x})$ is still symmetric by the same reasoning as in Proposition 4.4.2. We can derive an implicit formula for these new Schur functions in terms of the old ones by introducing another set of indeterminates $\mathbf{y} = (y_1, y_2, \ldots)$.

Proposition 4.9.1 ([Mac 79]) *Define $s_\lambda(\mathbf{x}, \mathbf{y}) = s_\lambda(x_1, x_2, \ldots, y_1, y_2, \ldots)$.* *Then*

$$s_\lambda(\mathbf{x}, \mathbf{y}) = \sum_{\mu \subseteq \lambda} s_\mu(\mathbf{x}) s_{\lambda/\mu}(\mathbf{y}). \tag{4.27}$$

Proof. The function $s_\lambda(\mathbf{x}, \mathbf{y})$ enumerates semistandard fillings of the diagram λ with letters from the totally ordered alphabet

$$\{1 < 2 < 3 < \cdots < 1' < 2' < 3' < \cdots\}.$$

In any such tableau, the unprimed numbers (which are weighted by the x's) form a subtableau of shape μ in the upper left corner of λ, whereas the primed numbers (weighted by the y's) fill the remaining squares of λ/μ. The right-hand side of (4.27) is the generating function for this description of the relevant tableaux. ∎

Since $s_{\lambda/\mu}(\mathbf{x})$ is symmetric, we can express it as a linear combination of ordinary Schur functions. Some familiar coefficients will then appear.

Theorem 4.9.2 *If the $c_{\mu\nu}^\lambda$ are Littlewood-Richardson coefficients, where $|\mu| + |\nu| = |\lambda|$, then*

$$s_{\lambda/\mu} = \sum_\nu c_{\mu\nu}^\lambda s_\nu.$$

Proof. Bring in yet a third set of variables $\mathbf{z} = \{z_1, z_2, \ldots\}$. By using the previous proposition and Cauchy's formula (Theorem 4.8.4),

$$
\begin{aligned}
\sum_{\lambda,\mu} s_\mu(\mathbf{x}) s_{\lambda/\mu}(\mathbf{y}) s_\lambda(\mathbf{z}) &= \sum_\lambda s_\lambda(\mathbf{x},\mathbf{y}) s_\lambda(\mathbf{z}) \\
&= \prod_{i,j} \frac{1}{1-x_i z_j} \frac{1}{1-y_i z_j} \\
&= \left[\sum_\mu s_\mu(\mathbf{x}) s_\mu(\mathbf{z}) \right] \left[\sum_\nu s_\nu(\mathbf{y}) s_\nu(\mathbf{z}) \right] \\
&= \sum_{\mu,\nu} s_\mu(\mathbf{x}) s_\nu(\mathbf{y}) s_\mu(\mathbf{z}) s_\nu(\mathbf{z}).
\end{aligned}
$$

Taking the coefficient of $s_\mu(\mathbf{x}) s_\nu(\mathbf{y}) s_\lambda(\mathbf{z})$ on both sides and comparing with equation (4.26) completes the proof. ∎

One last definition is needed to explain what the $c^\lambda_{\mu\nu}$ count.

Definition 4.9.3 A *ballot sequence*, or *lattice permutation*, is a sequence of positive integers $\pi = i_1 i_2 \ldots i_n$ such that, for any prefix $\pi_k = i_1 i_2 \ldots i_k$ and any positive integer l, the number of l's in π_k is at least as large as the number of $(l+1)$'s in that prefix. A *reverse* ballot sequence or lattice permutation is a sequence π such that π^r is a ballot sequence. ∎

As an example,

$$\pi = 1\ 1\ 2\ 3\ 2\ 1\ 3$$

is a lattice permutation, whereas

$$\pi = 1\ 2\ 3\ 2\ 1\ 1\ 3$$

is not because the prefix 1 2 3 2 has more twos than ones.

The name *ballot sequence* comes from the following scenario. Suppose the ballots from an election are being counted sequentially. Then a ballot sequence corresponds to a counting where candidate one always (weakly) leads candidate two, candidate two always (weakly) leads candidate three, and so on.

Furthermore, lattice permutations are just another way of encoding standard tableaux. Given P standard with n elements, form the sequence $\pi = i_1 i_2 \ldots i_n$, where $i_k = i$ if k appears in row i of P. The fact that the entries less than or equal to k form a partition with weakly decreasing parts translates to the ballot condition on π_k. It is easy to construct the inverse correspondence and see that our example lattice permutation codes the tableau

$$
P = \begin{array}{ccc} 1 & 2 & 6 \\ 3 & 5 & \\ 4 & 7 & \end{array} \ .
$$

We are finally ready to state and prove the Littlewood-Richardson rule. Although it was first stated by these two authors [L-R 34], complete proofs were not published until comparatively recently by Thomas [Tho 74, Tho 78] and Schützenberger [Scü 76]; our demonstration is based on the latter.

Theorem 4.9.4 (Littlewood-Richardson Rule [L-R 34]) *The value of the coefficient $c_{\mu\nu}^{\lambda}$ is equal to the number of semistandard tableaux T such that*

1. *T has shape λ/μ and content ν,*

2. *the row word of T, π_T, is a reverse lattice permutation.*

Proof. Let $d_{\mu\nu}^{\lambda}$ be the number of tableaux T satisfying the restrictions of the theorem. Then we claim that it suffices to find a weight-preserving map

$$T \xrightarrow{\;j\;} U \tag{4.28}$$

from semistandard tableaux T of shape λ/μ to semistandard tableaux U of normal shape such that

1. the number of tableaux T mapping to a given U of shape ν depends only on λ, μ, and ν and not on U itself;

2. the number of tableaux T mapping to a particular choice of U is $d_{\mu\nu}^{\lambda}$.

If such a bijection exists, then

$$s_{\lambda/\mu}(\mathbf{x}) = \sum_{\text{sh}\,T=\lambda/\mu} \mathbf{x}^T = \sum_{\nu} d_{\mu\nu}^{\lambda} \sum_{\text{sh}\,U=\nu} \mathbf{x}^U = \sum_{\nu} d_{\mu\nu}^{\lambda} s_{\nu}(\mathbf{x}).$$

Comparison of this expression with Theorem 4.9.2 and the fact that the Schur functions are a basis completes the proof.

It turns out that the map j needed for equation (4.28) is just the jeu de taquin! To show that j satisfies property 1, consider U and U' of the same shape ν. We will define a bijection between the set of tableaux T that map to U and the set of T' that map to U' as follows. Let P be a fixed standard tableau of shape μ. Then, in the notation of Section 3.8, we always have $U = j^P(T)$. Let the vacating tableau for this sequence of slides be denoted by Q_T, a standard tableau of shape λ/ν. Consider the composition

$$T \xrightarrow{\;j^P\;} U \longrightarrow U' \xrightarrow{\;j_{Q_T}\;} T', \tag{4.29}$$

i.e., slide T to normal shape to obtain U and Q_T, replace U by U', and then slide U' back out again according to the vacating tableau.

We first verify that this function is well defined in that T' must be sent to U' by jeu de taquin and must have shape λ/μ. The first statement is clear, since all the slides in j_{Q_T} are reversible. For the second assertion, note that by the definition of the vacating tableau,

$$T = j_{Q_T} j^P(T) = j_{Q_T}(U).$$

Also,

$$T' = j_{Q_T}(U').$$

Since U and U' are of the same normal shape, they are dual equivalent (Theorem 4.8.12), which implies that

$$T \stackrel{*}{\cong} T'. \qquad (4.30)$$

In particular, T and T' must have the same shape, as desired.

To show that (4.29) is a bijection, we claim that

$$T' \xrightarrow{j^P} U' \longrightarrow U \xrightarrow{j_{Q_{T'}}} T$$

is its inverse, where $Q_{T'} = v^P(T')$. But equation (4.30) implies that the corresponding vacating tableaux Q_T and $Q_{T'}$ are equal. So

$$j^P(T') = j^P j_{Q_T}(U') = j^P j_{Q_{T'}}(U') = U'$$

and

$$j_{Q_{T'}}(U) = j_{Q_{T'}} j^P(T) = j_{Q_T} j^P(T) = T,$$

showing that our map is a well-defined inverse.

Now we must choose a particular tableau U_0 for property 2. Let U_0 be the superstandard tableau of shape and content ν, where the elements of the ith row are all i's. Notice that the row word π_{U_0} is a reverse lattice permutation. In considering all tableaux mapped to U_0 by j, it is useful to have the following result.

Lemma 4.9.5 *If we can go from T to T' by a sequence of slides, then π_T is a reverse lattice permutation if and only if $\pi_{T'}$ is.*

Proof. It suffices to consider the case where T and T' differ by a single move. If the move is horizontal, the row word doesn't change, so consider a vertical move. Suppose

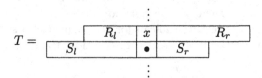

and

$$T' = \quad$$

where R_l and S_l (respectively, R_r and S_r) are the left (respectively, right) portions of the two rows between which x is moving. We show that if $\sigma = \pi_T^r$

is a ballot sequence, then so is $\sigma' = \pi_{T'}^r$. (The proof of the reverse implication is similar.)

Clearly, we need only check that the number of $(x + 1)$'s does not exceed the number of x's in any prefix of σ' ending with an element of R_l or S_r. To do this, we show that each $x + 1$ in such a prefix can be injectively matched with an x that comes before it. If the $x+1$ occurs in R_r or a higher row, then it can be matched with an x because σ is a ballot sequence. Notice that all these x's must be in higher rows because the x's in R_r come after the $(x+1)$'s in R_r when listed in the reverse row word. By semistandardness, there are no $(x+1)$'s in R_l to be matched. For the same reason, every $x + 1$ in S_r must have an x in R_r just above it. Since these x's have not been previously used in our matching, we are done. ∎

Now note that U_0 is the unique semistandard tableau of normal shape ν whose row word is a reverse lattice permutation. Thus by the lemma, the set of tableaux T with shape λ/μ such that $j(T) = U_0$ are precisely those described by the coefficients $d_{\mu\nu}^\lambda$. This finishes the proof of the Littlewood-Richardson rule. ∎

As an illustration, we can calculate the product $s_{(2,1)}s_{(2,2)}$. Listing all tableaux subject to the ballot sequence condition with content $(2,2)$ and skew shape $\lambda/(2,1)$ for some λ yields

$$
\begin{array}{llllll}
\bullet\ \bullet\ 1\ 1, & \bullet\ \bullet\ 1\ 1, & \bullet\ \bullet\ 1, & \bullet\ \bullet\ 1, & \bullet\ \bullet\ 1, & \bullet\ \bullet\ . \\
\bullet\ 2\ 2 & \bullet\ 2 & \bullet\ 1\ 2 & \bullet\ 1 & \bullet\ 2 & \bullet\ 1 \\
& 2 & 2 & 2\ 2 & 1 & 1\ 2 \\
& & & & 2 & 2
\end{array}
$$

Thus

$$s_{(2,1)}s_{(2,2)} = s_{(4,3)} + s_{(4,2,1)} + s_{(3^2,1)} + s_{(3,2^2)} + s_{(3,2,1^2)} + s_{(2^3,1)}.$$

For another example, let us find the coefficient of $s_{(5,3,2,1)}$ in $s_{(3,2,1)}s_{(3,2)}$. First we fix the outer shape $\lambda = (5,3,2,1)$ and then find the tableaux as before:

$$
\begin{array}{lll}
\bullet\ \bullet\ \bullet\ 1\ 1, & \bullet\ \bullet\ \bullet\ 1\ 1, & \bullet\ \bullet\ \bullet\ 1\ 1. \\
\bullet\ \bullet\ 1 & \bullet\ \bullet\ 2 & \bullet\ \bullet\ 2 \\
\bullet\ 2 & \bullet\ 1 & \bullet\ 2 \\
2 & 2 & 1
\end{array}
$$

It follows that

$$c_{(3,2,1)(3,2)}^{(5,3,2,1)} = 3.$$

4.10 The Murnaghan-Nakayama Rule

The Murnaghan-Nakayama rule [Mur 37, Nak 40] is a combinatorial way of computing the value of the irreducible character χ^λ on the conjugacy class α. The crucial objects that come into play are the skew hooks.

Definition 4.10.1 A *skew hook*, or *rim hook*, ξ, is a skew diagram obtained by taking all cells on a finite lattice path with steps one unit northward or eastward. Equivalently, ξ is a skew hook if it is edgewise connected and contains no 2×2 subset of cells:

$$\qquad\qquad (4.31)$$

Generalizing the definition in Section 3.10, the *leg length* of ξ is

$$ll(\xi) = \text{(the number of rows of ξ)} - 1. \blacksquare$$

For example,

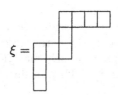

has $ll(\xi) = 4$. The name *rim hook* comes from the fact that such a diagram can be obtained by projecting a regular hook along diagonals onto the boundary of a shape. Our example hook is the projection of $H_{1,1}$ onto the rim of $\lambda = (6, 3, 3, 1, 1)$.

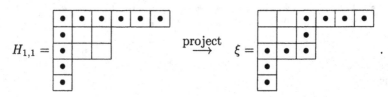

Notice that we are using single Greek letters such as ξ for rim hooks even though they are skew. If $\xi = \lambda/\mu$, then we write $\lambda\backslash\xi$ for μ. In the preceding example, $\mu = (2, 2)$. Also, if $\alpha = (\alpha_1, \alpha_2, \ldots, \alpha_k)$ is a composition, then let $\alpha\backslash\alpha_1 = (\alpha_2, \ldots, \alpha_k)$. With this notation, we can state the main result of this section.

Theorem 4.10.2 (Murnaghan-Nakayama Rule [Mur 37, Nak 40]) *If λ is a partition of n and $\alpha = (\alpha_1, \ldots, \alpha_k)$ is a composition of n, then we have*

$$\chi^\lambda_\alpha = \sum_\xi (-1)^{ll(\xi)} \chi^{\lambda\backslash\xi}_{\alpha\backslash\alpha_1},$$

where the sum runs over all rim hooks ξ of λ having α_1 cells.

Before proving this theorem, a few remarks are in order. To calculate χ^λ_α the rule must be used iteratively. First remove a rim hook from λ with α_1 cells in all possible ways such that what is left is a normal shape. Then strip away hooks with α_2 squares from the resulting diagrams, and so on. At some stage either it will be impossible to remove a rim hook of the right size (so the contribution of the corresponding character is zero), or all cells will be deleted (giving a contribution of $\pm\chi_{(0)}^{(0)} = \pm 1$).

To illustrate the process, we compute $\chi_{(5,4,2)}^{(4,4,3)}$. The stripping of hooks can be viewed as a tree. Cells to be removed are marked with a dot, and the appropriate sign appears to the right of the diagram:

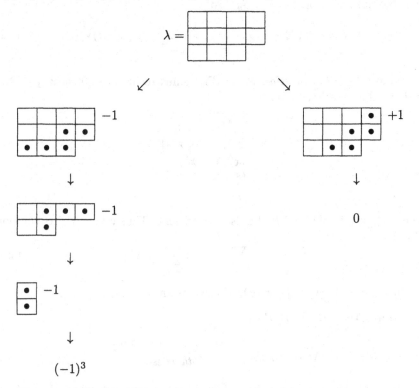

The corresponding calculations are

$$\begin{aligned}
\chi_{(5,4,2)}^{(4,4,3)} &= -\chi_{(4,2)}^{(4,2)} + \chi_{(4,2)}^{(3,2,1)} \\
&= -(-\chi_{(2)}^{(1,1)}) + 0 \\
&= -(-(-\chi_{(0)}^{(0)})) \\
&= (-1)^3.
\end{aligned}$$

Also note that the branching rule from Section 2.8 is a special case of Murnaghan-Nakayama. Take $\alpha = (1, \alpha_2, \ldots, \alpha_k)$ and let $\pi \in S_n$ have type α. Since π has a fixed-point,

$$\chi_\alpha^\lambda = \chi^\lambda(\pi) = \chi^\lambda\!\downarrow_{S_{n-1}} (\pi),$$

which corresponds to the left-hand side of the first equation in Theorem 2.8.3. As for the right side, $|\xi| = 1$ forces $\lambda\backslash\xi$ to be of the form λ^- with all signs $(-1)^0 = +1$.

We now embark on a proof of the Murnaghan-Nakayama rule. It is based on the one in James [Jam 78, pp. 79–83].

Proof (of Theorem 4.10.2). Let $m = \alpha_1$. Consider $\pi\sigma \in \mathcal{S}_{n-m} \times \mathcal{S}_m \subseteq \mathcal{S}_n$, where π has type $(\alpha_2, \ldots, \alpha_k)$ and σ is an m-cycle. By part 2 of Theorem 1.11.3, the characters $\chi^\mu \otimes \chi^\nu$, where $\mu \vdash n - m$, $\nu \vdash m$, form a basis for the class functions on $\mathcal{S}_{n-m} \times \mathcal{S}_m$. So

$$\chi^\lambda_\alpha = \chi^\lambda(\pi\sigma) = \chi^\lambda\!\downarrow_{\mathcal{S}_{n-m}\times\mathcal{S}_m}(\pi\sigma) = \sum_{\substack{\mu \vdash n-m \\ \nu \vdash m}} m^\lambda_{\mu\nu}\chi^\mu(\pi)\chi^\nu(\sigma). \qquad (4.32)$$

To find the multiplicities, we use Frobenius reciprocity (Theorem 1.12.6) and the characteristic map:

$$
\begin{aligned}
m^\lambda_{\mu\nu} &= \langle \chi^\lambda\!\downarrow_{\mathcal{S}_{n-m}\times\mathcal{S}_m}, \chi^\mu \otimes \chi^\nu \rangle \\
&= \langle \chi^\lambda, (\chi^\mu \otimes \chi^\nu)\!\uparrow^{\mathcal{S}_n} \rangle \\
&= \langle \chi^\lambda, \chi^\mu \cdot \chi^\nu \rangle \\
&= \langle s_\lambda, s_\mu s_\nu \rangle \\
&= c^\lambda_{\mu\nu},
\end{aligned}
$$

where $c^\lambda_{\mu\nu}$ is a Littlewood-Richardson coefficient. Thus we can write equation (4.32) as

$$\chi^\lambda(\pi\sigma) = \sum_{\mu \vdash n-m} \chi^\mu(\pi) \sum_{\nu \vdash m} c^\lambda_{\mu\nu}\chi^\nu(\sigma). \qquad (4.33)$$

Now we must evaluate $\chi^\nu(\sigma)$, where σ is an m-cycle.

Lemma 4.10.3 *If $\nu \vdash m$, then*

$$\chi^\nu_{(m)} = \begin{cases} (-1)^{m-r} & \text{if } \nu = (r, 1^{m-r}), \\ 0 & \text{otherwise.} \end{cases}$$

Note that this is a special case of the theorem we are proving. If $\alpha = (m)$, then $\chi^\nu_\alpha \neq 0$ only when we can remove all the cells of ν in a single sweep—that is, when ν itself is a hook diagram $(r, 1^{m-r})$. In this case, the Murnaghan-Nakayama sum has a single term with leg length $m - r$.

Proof (of Lemma 4.10.3). By equation (4.23), $\chi^\nu_{(m)}$ is $z_{(m)} = m$ times the coefficient of p_m in

$$s_\nu = \sum_\mu \frac{1}{z_\mu}\chi^\nu_\mu p_\mu.$$

Using the complete homogeneous Jacobi-Trudi determinant (Theorem 4.5.1), we obtain

$$s_\nu = |h_{\nu_i - i + j}|_{l \times l} = \sum_\kappa \pm h_\kappa,$$

where the sum is over all compositions $\kappa = (\kappa_1, \ldots, \kappa_l)$ that occur as a term in the determinant. But each h_{κ_i} in h_κ can be written as a linear combination of power sums. So, since the p's are a multiplicative basis, the resulting linear combination for h_κ will not contain p_m unless κ contains exactly one nonzero part, which must, of course, be m. Hence $\chi^\nu_{(m)} \neq 0$ only when h_m appears in the preceding determinant.

The largest index to appear in this determinant is at the end of the first row, and $\nu_1 - 1 + l = h_{1,1}$, the hooklength of cell $(1,1)$. Furthermore, we always have $m = |\nu| \geq h_{1,1}$. Thus $\chi^\nu_{(m)}$ is nonzero only when $h_{1,1} = m$, i.e., when ν is a hook $(r, 1^{m-r})$. In this case, we have

$$
s_\nu = \begin{vmatrix}
h_r & \cdots & & & & h_m \\
h_0 & h_1 & \cdots & & & \\
0 & h_0 & h_1 & \cdots & & \\
0 & 0 & h_0 & h_1 & \cdots & \\
\vdots & \vdots & \vdots & \vdots & \vdots & \vdots
\end{vmatrix}
$$

$$
= (-1)^{m-r} h_m + \text{other terms not involving } p_m.
$$

But $h_m = s_{(m)}$ corresponds to the trivial character, so comparing coefficients of p_m/m in this last set of equalities yields $\chi^\nu_{(m)} = (-1)^{m-r}$, as desired. ∎

In view of this lemma and equation (4.33), we need to find $c^\lambda_{\mu\nu}$ when ν is a hook. This turns out to just be a binomial coefficient.

Lemma 4.10.4 *Let $\lambda \vdash n$, $\mu \vdash n - m$, and $\nu = (r, 1^{m-r})$. Then $c^\lambda_{\mu\nu} = 0$ unless each edgewise connected component of λ/μ is a rim hook. In that case, if there are k component hooks spanning a total of c columns, then*

$$
c^\lambda_{\mu\nu} = \binom{k-1}{c-r}.
$$

Proof. By the Littlewood-Richardson rule (Theorem 4.9.4), $c^\lambda_{\mu\nu}$ is the number of semistandard tableaux T of shape λ/μ containing r ones and a single copy each of $2, 3, \ldots, m-r+1$ such that π_T is a reverse lattice permutation. Thus the numbers greater than one in π^r_T must occur in increasing order. This condition, together with semistandardness, puts the following constraints on T:

T1. Any cell of T having a cell to its right must contain a one.

T2. Any cell of T having a cell above must contain an element bigger than one.

If T contains a 2×2 block of squares, as in (4.31), then there is no way to fill the lower left cell and satisfy both T1 and T2. Thus $c^\lambda_{\mu\nu} = 0$ if the components of the shape of T are not rim hooks.

Now suppose $\lambda/\mu = \biguplus_{i=1}^{k} \xi^{(i)}$, where each $\xi^{(i)}$ is a component skew hook. Conditions T1, T2 and the fact that 2 through $m-r+1$ increase in π^r_T show that every rim hook must have the form

		1	1	1	b
		d			
$\xi^{(i)} =$	1	1	$d+1$		
$d+2$					
$d+3$					

where $d > 1$ is the smallest number that has not yet appeared in π_T^r and b is either 1 or $d - 1$. Thus all the entries in $\xi^{(i)}$ are determined once we choose the value of b. Furthermore, in $\xi^{(1)}$ we must have $b = 1$. By T2 we have $c - (k - 1)$ ones already fixed in T. Hence there are $r - c + k - 1$ ones left to distribute among the $k - 1$ cells marked with a b. The number of ways this can be done is

$$c_{\mu\nu}^{\lambda} = \binom{k-1}{r-c+k-1} = \binom{k-1}{c-r}. \ \blacksquare$$

Putting the values from the two lemmas in equation (4.33) we obtain

$$\chi^{\lambda}(\pi\sigma) = \sum_{\mu} \chi^{\mu}(\pi) \sum_{r=1}^{m} \binom{k-1}{c-r} (-1)^{m-r}. \tag{4.34}$$

Now, $k \leq c \leq m$, since these three quantities represent the number of skew hooks $\xi^{(i)}$, the number of columns in the $\xi^{(i)}$, and the number of cells in the $\xi^{(i)}$, respectively. Thus, using Exercise 1d of this chapter,

$$\sum_{r=1}^{m} \binom{k-1}{c-r} (-1)^{m-r} = (-1)^{m-c} \left[\binom{k-1}{0} - \binom{k-1}{1} + \cdots \pm \binom{k-1}{k-1} \right]$$

$$= \begin{cases} (-1)^{m-c} & \text{if } k-1=0, \\ 0 & \text{otherwise.} \end{cases}$$

But if $k = 1$, then λ/μ is a single skew hook ξ with m squares and c columns. Hence $m - c = ll(\xi)$, so equation (4.34) becomes

$$\chi^{\lambda}(\pi\sigma) = \sum_{|\xi|=m} (-1)^{ll(\xi)} \chi^{\lambda\backslash\xi}(\pi),$$

finishing the proof of the Murnaghan-Nakayama rule. $\blacksquare$

It is possible to combine the calculations for χ_{α}^{λ} into individual tableaux.

Definition 4.10.5 A *rim hook tableau* is a generalized tableau T with positive integral entries such that

1. rows and columns of T weakly increase, and

2. all occurrences of i in T lie in a single rim hook. $\blacksquare$

Obviously, the stripping of rim hooks from a diagram determines a rim hook tableau by labeling the last hook removed with ones, the next to last with twos, and so on. For example, the tableau corresponding to the left branch of the preceding tree is

$$T = \begin{array}{cccc} 1 & 2 & 2 & 2 \\ 1 & 2 & 3 & 3 \\ 3 & 3 & 3 & \end{array}.$$

Now define the *sign* of a rim hook tableau with rim hooks $\xi^{(i)}$ to be

$$(-1)^T = \prod_{\xi^{(i)} \in T} (-1)^{ll(\xi^{(i)})}.$$

The tableau just given has $\operatorname{sgn}(T) = (-1)^3$. It is easy to see that the Murnaghan-Nakayama theorem can be restated.

Corollary 4.10.6 *Let λ be a partition of n and let $\alpha = (\alpha_1, \ldots, \alpha_k)$ be any composition of n. Then*

$$\chi^\lambda_\alpha = \sum_T (-1)^T,$$

where the sum is over all rim hook tableaux of shape λ and content α. ∎

It is worth mentioning that Stanton and White [S-W 85] have extended Schensted's construction to rim hook tableaux. This, in turn, has permitted White [Whi 83] to give a bijective proof of the orthonormality of the irreducible characters of $\mathcal{S}_n$ (Theorem 1.9.3) analogous to the combinatorial proof of

$$n! = \sum_{\lambda \vdash n} (f^\lambda)^2,$$

which is just the fact that the first column of the character table has unit length.

4.11 Exercises

1. Recall that the *binomial coefficients* are defined by

$$\binom{n}{k} = \text{the number of ways to choose } S \subseteq \{1, 2, \ldots, n\} \text{ with } |S| = k.$$

Note that this definition makes sense for all $n \geq 0$ and all integers k where $\binom{n}{k} = 0$ if $k < 0$ or $k > n$.

(a) Show that for $n \geq 1$ we have the recursion

$$\binom{n}{k} = \binom{n-1}{k} + \binom{n-1}{k-1}$$

with boundary condition $\binom{0}{k} = \delta_{k,0}$.

(b) Show that for $0 \le k \le n$ we have

$$\binom{n}{k} = \frac{n!}{k!(n-k)!}$$

in two different ways: inductively and by a direct counting argument.

(c) Show that

$$\sum_{k \ge 0} \binom{n}{k} x^k = (1+x)^n$$

in two different ways: inductively and by using weight generating functions.

(d) Show that

$$\sum_{k \ge 0} \binom{n}{k} = 2^n \quad \text{and} \quad \sum_{k \ge 0} (-1)^k \binom{n}{k} = \delta_{n,0},$$

where $\delta_{n,0}$ is the Kronecker delta, in two different ways: using the previous exercise and inductively.

(e) Let S be a set. Show that

$$\sum_{T \subseteq S} (-1)^{|T|} = \delta_{|S|,0}$$

in two different ways: using the previous exercise, and by assigning a sign to every $T \subseteq S$ and then constructing a fixed-point free, sign-reversing involution on the set of subsets of S.

(f) Prove *Vandermonde's convolution*

$$\sum_{i \ge 0} \binom{n}{i}\binom{m}{k-i} = \binom{n+m}{k}$$

in three different ways: inductively, using generating functions, and by a counting argument.

2. Prove the following theorem of Glaisher. The number of partitions of n with no part divisible by k is equal to the number of partitions of n with each part appearing at most $k - 1$ times. Which theorem in the text is a special case of this result?

3. Let $p(n, k)$ (respectively, $p_d(n, k)$) denote the number of partitions of n into k parts (respectively, k distinct parts). Show that

$$\sum_{\substack{n \ge 0 \\ k \ge 0}} p(n, k) x^n t^k = \prod_{i \ge 1} \frac{1}{1 - x^i t}$$

$$= \sum_{n \geq 0} \frac{x^n t^n}{(1-x)(1-x^2)\cdots(1-x^n)},$$

$$\sum_{\substack{n \geq 0 \\ k \geq 0}} p_d(n,k) x^n t^k = \prod_{i \geq 1}(1 + x^i t)$$

$$= \sum_{n \geq 0} \frac{x^{\binom{n+1}{2}} t^n}{(1-x)(1-x^2)\cdots(1-x^n)}.$$

4. Consider the plane $\mathcal{P} = \{(i,j) \, : \, i,j \geq 1\}$, i.e., the "shape" consisting of an infinite number of rows each of infinite length. A *plane partition of n* is an array, T, of nonnegative integers in $\mathcal{P}$ such that T is weakly decreasing in rows and columns and $\sum_{(i,j) \in \mathcal{P}} T_{i,j} = n$. For example, the plane partitions of 3 (omitting zeros) are

$$3, \quad 2\,1, \quad 2, \quad 1\,1\,1, \quad 1\,1, \quad 1.$$
$$ 1 1 1$$
$$ 1$$

(a) Let

$$pp(n) = \text{ the number of plane partitions of } n$$

and show that

$$\sum_{n \geq 0} pp(n) x^n = \prod_{i \geq 1} \frac{1}{(1-x^i)^i}$$

in two ways: combinatorially and by taking a limit.

(b) Define the *trace* of the array T to be $\sum_i T_{i,i}$. Let

$$pp(n,k) = \text{ the number of plane partitions of } n \text{ with trace } k$$

and show that

$$\sum_{n,k \geq 0} pp(n,k) x^n t^k = \prod_{i \geq 1} \frac{1}{(1-x^i t)^i}.$$

5. (a) Let τ be a rooted tree (see Chapter 3, Exercise 18). Show that the generating function for reverse τ-partitions is

$$\prod_{v \in \tau} \frac{1}{1 - x^{h_v}}$$

and use this to rederive the hook formula for rooted trees.

(b) Let λ^* be a shifted shape (see Chapter 3, Exercise 21). Show that the generating function for reverse λ^*-partitions is

$$\prod_{(i,j) \in \lambda^*} \frac{1}{1 - x^{h_{i,j}^*}}$$

and use this to rederive the hook formula for shifted tableaux.

6. The *principal specialization* of a symmetric function in the variables $\{x_1, x_2, \ldots, x_m\}$ is obtained by replacing x_i by q^i for all i.

   (a) Show that the Schur function specialization $s_\lambda(q, q^2, \ldots, q^m)$ is the generating function for semistandard λ-tableaux with all entries of size at most m.

   (b) Define the *content* of cell (i, j) to be

   $$c_{i,j} = j - i.$$

   Prove that

   $$s_\lambda(q, q^2, \ldots, q^m) = q^{m(\lambda)} \prod_{(i,j) \in \lambda} \frac{1 - q^{c_{i,j}+m}}{1 - q^{h_{i,j}}},$$

   where $m(\lambda) = \sum_{i \geq 1} i\lambda_i$. Use this result to rederive Theorem 4.2.2.

7. Prove part 2 of Theorem 4.3.7.

8. Let $\Lambda_{\mathbb{Z}}^n$ denote the ring of symmetric functions of degree n with integral coefficients. An *integral basis* is a subset of $\Lambda_{\mathbb{Z}}^n$ such that every element of $\Lambda_{\mathbb{Z}}^n$ can be uniquely expressed as an integral linear combination of basis elements. As λ varies over all partitions of n, which of the following are integral bases of $\Lambda_{\mathbb{Z}}^n$: m_λ, p_λ, e_λ, h_λ, s_λ?

9. (a) Let

   $$e_k(n) = e_k(x_1, x_2, \ldots, x_n)$$

   and similarly for the homogeneous symmetric functions. Show that we have the following recursion relations and boundary conditions. For $n \geq 1$,

   $$e_k(n) = e_k(n-1) + x_n e_{k-1}(n-1),$$
   $$h_k(n) = h_k(n-1) + x_n h_{k-1}(n),$$

   and $e_k(0) = h_k(0) = \delta_{k,0}$, where $\delta_{k,0}$ is the Kronecker delta.

   (b) The *(signless) Stirling numbers of the first kind* are defined by

   $$c(n, k) = \text{the number of } \pi \in S_n \text{ with } k \text{ disjoint cycles.}$$

   The *Stirling numbers of the second kind* are defined by

   $$S(n, k) = \text{the number of partitions of } \{1, \ldots, n\} \text{ into } k \text{ subsets.}$$

   Show that we have the following recursion relations and boundary conditions. For $n \geq 1$,

   $$\begin{aligned} c(n, k) &= c(n-1, k-1) + (n-1)c(n-1, k) \\ S(n, k) &= S(n-1, k-1) + kS(n-1, k) \end{aligned}$$

   and $c(0, k) = S(0, k) = \delta_{k,0}$.

(c) For the binomial coefficients and Stirling numbers, show that

$$\binom{n}{k} = e_k(\overbrace{1,1,\ldots,1}^{n})$$

$$= h_k(\overbrace{1,1,\ldots,1}^{n-k+1}),$$
$$c(n,k) = e_{n-k}(1,2,\ldots,n-1),$$
$$S(n,k) = h_{n-k}(1,2,\ldots,k).$$

10. Prove the elementary symmetric function version of the Jacobi-Trudi determinant.

11. A sequence $(a_k)_{k\geq0} = a_0, a_1, a_2, \ldots$ of real numbers is *log concave* if

$$a_{k-1}a_{k+1} \leq a_k^2 \qquad \text{for all } k > 0.$$

(a) Show that if $a_k > 0$ for all k, then log concavity is equivalent to the condition

$$a_{k-1}a_{l+1} \leq a_k a_l \qquad \text{for all } l \geq k > 0.$$

(b) Show that the following sequences of binomial coefficients and Stirling numbers (see Exercise 9) are log concave in two ways: inductively and using results from this chapter.

$$\left(\left(\binom{n}{k}\right)\right)_{k\geq0}, \qquad \left(\left(\binom{n}{k}\right)\right)_{n\geq0}, \qquad (c(n,k))_{k\geq0}, \qquad (S(n,k))_{n\geq0}.$$

(c) Show that if $(x_n)_{n\geq0}$ is a log concave sequence of positive reals then, for fixed k, the sequences

$$(e_k(x_1,\ldots,x_n))_{n\geq0} \quad \text{and} \quad (h_k(x_1,\ldots,x_n))_{n\geq0}$$

are also log concave. Conclude that the following sequences are log concave for fixed n and k:

$$(c(n+j,k+j))_{j\geq0}, \qquad (S(n+j,k+j))_{j\geq0}.$$

12. (a) Any two bases $\{u_\lambda : \lambda \vdash n\}$ and $\{v_\lambda : \lambda \vdash n\}$ for Λ^n define a unique inner product by

$$\langle u_\lambda, v_\mu \rangle = \delta_{\lambda\mu}$$

and sesquilinear extension. Also, these two bases define a generating function

$$f(\mathbf{x},\mathbf{y}) = \sum_{\lambda\vdash n} u_\lambda(\mathbf{x})\overline{v_\lambda(\mathbf{y})}.$$

Prove that two pairs of bases define the same inner product if and only if they define the same generating function.

(b) Show that
$$\sum_\lambda \frac{1}{z_\lambda} p_\lambda(\mathbf{x}) p_\lambda(\mathbf{y}) = \prod_{i,j} \frac{1}{1 - x_i y_j}.$$

13. Prove the following. Compare with Exercise 7 in Chapter 3.

   (a) Theorem 4.8.7

   (b) There is a bijection $M \longleftrightarrow T$ between symmetric matrices $M \in$ Mat and semistandard tableaux such that the trace of M is the number of columns of odd length of T.

   (c) We have the following equalities:
   $$\sum_\lambda s_\lambda = \prod_i \frac{1}{1 - x_i} \prod_{i<j} \frac{1}{1 - x_i x_j},$$
   $$\sum_{\lambda' \text{even}} s_\lambda = \prod_{i<j} \frac{1}{1 - x_i x_j}.$$

14. Give a second proof of the Littlewood-Richardson rule by finding a bijection
   $$(P, Q) \longleftrightarrow (T, U)$$
   such that

   - P, Q, T, and U are semistandard of shape μ, ν, λ, and λ/μ, respectively,

   - $\text{cont}\,P + \text{cont}\,Q = \text{cont}\,T$, where compositions add componentwise, and

   - π_U is a reverse lattice permutation of type ν.

15. The *Durfee size* of a partition λ is the number of diagonal squares (i, i) in its shape. Show that $\chi_\alpha^\lambda = 0$ if the Durfee size of λ is bigger than the number of parts of α.

Chapter 5

Applications and Generalizations

In this chapter we will give some interesting applications and extensions of results found in the rest of the book. Before tackling this material, the reader may find it helpful to review the poset definitions and examples at the beginning of Section 2.2 and Definition 2.5.7.

We begin with a couple of sections studying differential posets, which were introduced by Stanley [Stn 88]. This permits us to give two more proofs of the sum-of-squares formula (5.1). One uses linear algebra and the other, due to Fomin [Fom 86], is bijective. But both of them generalize from the poset called Young's lattice (the classical case) to the wider class of differential posets.

Next comes a pair of sections concerning groups acting on posets [Stn 82]. In the first we study some interesting modules that arise from the symmetric group acting on the Boolean algebra. The second shows how group actions can be used to prove unimodality results.

The final section introduces graph colorings. In particular, it shows how certain symmetric functions of Stanley [Stn 95] can be used to study such objects.

5.1 Young's Lattice and Differential Posets

We are going to give a third proof of the formula

$$\sum_{\lambda \vdash n} (f^\lambda)^2 = n! \tag{5.1}$$

based on properties of a poset called Young's lattice. Two interesting points about this demonstration are that it uses linear algebra and generalizes to the larger family of differential partially ordered sets.

191

Definition 5.1.1 Let $(A, \leq)$ be a partially ordered set. Then $a, b \in A$ have a *greatest lower bound*, or *meet* ,if there is $c \in A$ such that

$$1.\ c \leq a,\ c \leq b \quad \text{and} \quad 2.\ \text{if } d \leq a,\ d \leq b, \text{ then } d \leq c. \qquad (5.2)$$

The meet of a and b, if it exists, is denoted by $a \wedge b$. The *least upper bound*, or *join*, of a and b is defined by reversing all the inequalities in (5.2) and is written $a \vee b$. A poset where every pair of elements has a meet and a join is called a *lattice*. ∎

As examples, the Boolean algebra B_n is a lattice, since if $S, T \in B_n$ then

$$S \wedge T = S \cap T \quad \text{and} \quad S \vee T = S \cup T$$

This relationship between $\wedge$ and $\cap$, as well as $\vee$ and $\cup$, makes the symbols for meet and join easy to remember. The n-chain C_n is also a lattice, since if $i, j \in C_n$ then

$$i \wedge j = \min\{i, j\} \quad \text{and} \quad i \vee j = \max\{i, j\}.$$

The lattice that concerns us here was first studied by Kreweras [Kre 65].

Definition 5.1.2 *Young's lattice*, Y, is the poset of all integer partitions ordered by inclusion of their Ferrers diagrams. ∎

Here is a picture of the bottom portion of Y (since it is infinite, it would be hard to draw the whole thing!) :

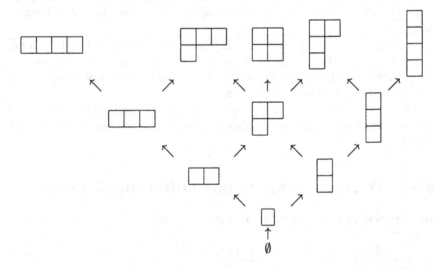

It is easy to verify that Y is a lattice with $\wedge$ and $\vee$ being intersection and union of Ferrers diagrams. Note that Y has a minimum, namely the empty partition $\emptyset$ of 0, but no maximum. Here are a couple of other basic properties of Young's lattice.

Proposition 5.1.3 *The poset Y satisfies the following two conditions.*

(1) If $\lambda \in Y$ covers k elements for some k, then it is covered by $k + 1$ elements.

(2) If $\lambda \neq \mu$ and λ and μ both cover l elements for some l, then they are both covered by l elements. In fact, $l \leq 1$.

Proof. (1) The elements covered by λ are just the partitions λ^- obtained by removing an inner corner of λ, while those covering λ are the λ^+ obtained by adding an outer corner. But along the rim of λ, inner and outer corners alternate beginning and ending with an outer corner. So the number of λ^+ is one greater than the number of λ^-.

(2) If λ, μ both cover an element of Y, then this element must be $\lambda \wedge \mu$. So $l \leq 1$. Also, $l = 1$ if and only if $|\lambda| = |\mu| = n$ and $|\lambda \cap \mu| = n - 1$ for some n. But this is equivalent to $|\lambda| = |\mu| = n$ and $|\lambda \cup \mu| = n + 1$, i.e., λ, μ are both covered by a single element. ∎

Note that each of the implications in this proposition is equivalent to its converse, so they could actually have been stated as "if and only if." We can rephrase these two conditions in terms of linear operators.

Definition 5.1.4 Consider the vector space $\mathbb{C}Y$ of formal linear combinations of elements of Y. Define two linear operators, *down*, D, and *up*, U, by

$$D(\lambda) = \sum_{\lambda^- \lessdot \lambda} \lambda^-,$$

$$U(\lambda) = \sum_{\lambda^+ \gtrdot \lambda} \lambda^+,$$

and linear extension. ∎

For example

$$D\left(\begin{smallmatrix} \bullet & \bullet & \bullet \\ \bullet & & \\ \bullet & & \end{smallmatrix} \right) = \begin{smallmatrix} \bullet & \bullet & \\ \bullet & & \\ \bullet & & \end{smallmatrix} + \begin{smallmatrix} \bullet & \bullet & \bullet \\ \bullet & & \end{smallmatrix} ,$$

$$U\left(\begin{smallmatrix} \bullet & \bullet & \bullet \\ \bullet & & \\ \bullet & & \end{smallmatrix} \right) = \begin{smallmatrix} \bullet & \bullet & \bullet & \bullet \\ \bullet & & & \\ \bullet & & & \end{smallmatrix} + \begin{smallmatrix} \bullet & \bullet & \bullet \\ \bullet & \bullet & \\ \bullet & & \end{smallmatrix} + \begin{smallmatrix} \bullet & \bullet & \bullet \\ \bullet & & \\ \bullet & & \\ \bullet & & \end{smallmatrix} .$$

Proposition 5.1.5 *The operators D, U satisfy*

$$DU - UD = I, \tag{5.3}$$

where I is the identity map.

Proof. By linearity, it suffices to prove that this equation holds when applied to a single $\lambda \in \mathbb{C}Y$. But by the previous proposition

$$DU(\lambda) = \sum_{\mu} \mu + (k+1)\lambda,$$

where the sum is over all $\mu \neq \lambda$ such that there is an element of Y covering both, and $k+1$ is the number of elements covering λ. In the same way,

$$UD(\lambda) = \sum_{\mu} \mu + k\lambda,$$

and taking the difference completes the proof. ■

Equation (5.3) is a special case of the next result.

Corollary 5.1.6 *Let $p(x) \in \mathbb{C}[x]$ be a polynomial. Then*

$$Dp(U) = p'(U) + p(U)D,$$

where $p'(x)$ is the derivative.

Proof. By linearity, it suffices to prove this identity for the powers U^n, $n \geq 0$. The case $n = 0$ is trivial. So, applying induction yields

$$
\begin{aligned}
DU^{n+1} &= (DU^n)U \\
&= (nU^{n-1} + U^n D)U \\
&= nU^n + U^n(I + UD) \\
&= (n+1)U^n + U^{n+1}D. \quad \blacksquare
\end{aligned}
$$

It is also possible to obtain tableaux from Young's lattice using chains.

Definition 5.1.7 *Let A be a poset and consider $a \leq b$ in A. An a–b (ascending) chain of length n is a sequence*

$$C : a = a_0 < a_1 < a_1 < \cdots < a_n = b. \tag{5.4}$$

A descending chain is defined analogously by reversing all the inequalities. An a–b chain is saturated if it is contained in no larger a–b chain. ■

So a chain of length n is just a copy of C_n in A. In Y, a chain is saturated if and only if each of the inequalities in (5.4) is a cover. Also, there is a simple bijection between standard λ-tableaux P and saturated $\emptyset$–λ chains $\emptyset = \lambda^0 \prec \lambda^1 \prec \cdots \prec \lambda^n = \lambda$ given by

$$\lambda^i = \text{shape of the tableau of elements} \leq i \text{ in } P.$$

(This is essentially the same as the way we converted a tabloid into a sequence of compositions in Section 2.5.) For example, the tableau

$$P = \begin{array}{cc} 1 & 3 \\ 2 & 5 \\ 4 & \end{array}$$

corresponds to the saturated chain

One could just as well put saturated descending chains in bijection with tableaux, or in terms of the D and U operators and partitions $\lambda \vdash n$,

$$
\begin{aligned}
D^n(\lambda) &= f^\lambda \emptyset, \\
U^n(\emptyset) &= \sum_{\lambda \vdash n} f^\lambda \lambda.
\end{aligned}
$$

Also, a pair of tableaux of the same shape λ can be viewed as a saturated $\emptyset$–λ chain followed by one from λ back to $\emptyset$. So

$$
D^n U^u(\emptyset) = \emptyset \sum_{\lambda \vdash n} (f^\lambda)^2. \tag{5.5}
$$

We now have all the ingredients to give another proof of the sum-of-squares formula.

Theorem 5.1.8 *We have*
$$
\sum_{\lambda \vdash n} (f^\lambda)^2 = n!
$$

Proof. In view of equation (5.5) we need only show that $D^n U^n(\emptyset) = n!\emptyset$. We will do this by induction, it being clear for $n = 0$. But using Corollary 5.1.6 and the fact that $\emptyset$ is the minimal element of Y, we obtain

$$
\begin{aligned}
D^n U^n(\emptyset) &= D^{n-1}(DU^n)(\emptyset) \\
&= D^{n-1}(nU^{n-1} + U^n D)(\emptyset) \\
&= n D^{n-1} U^{n-1}(\emptyset) + D^{n-1} U^n D(\emptyset) \\
&= n(n-1)!\emptyset + 0. \quad\blacksquare
\end{aligned}
$$

It turns out that all these results for Young's lattice can be generalized to a wider family of posets, as has been done by Stanley [Stn 88, Stn 90]. (See also Fomin [Fom 94].) First, however, we need a few more definitions.

Definition 5.1.9 Let A be a poset having a minimum element, $\hat{0}$. Then A is *graded* if, for each $a \in A$, all $\hat{0}$–a saturated chains have the same length. We denote this length by rk a, called the *rank* of a. Furthermore, the *kth rank of* A is the set

$$
A_k = \{a \in A : \operatorname{rk} a = k\}.
$$

Finally, we say that A *has rank* n if there is a nonnegative integer n such that

$$
n = \max\{\operatorname{rk} a : a \in A\}. \quad\blacksquare
$$

The posets C_n, B_n, and Y are all graded. Their kth ranks are the number k, all subsets of $\{1, \ldots, n\}$ with k elements, and all partitions of k, respectively. The first two posets have rank n, but Y has arbitrarily long chains and so has no rank.

Definition 5.1.10 ([Stn 88]) A poset A is *differential* if it satisfies the following three conditions.

DP1 A has a (unique) minimum element $\hat{0}$ and is graded.

DP2 If $a \in A$ covers k elements for some k, then it is covered by $k + 1$ elements.

DP3 If $a \neq b$ and a and b both cover l elements for some l, then they are both covered by l elements. ∎

Actually, these posets are called 1-differential by Stanley. For the definition of his more general r-differential posets, see Exercise 6. By Proposition 5.1.3, Y is a differential poset. For another example, see Exercise 2.

As with Y, we can show that the integer l in DP3 is at most 1.

Lemma 5.1.11 *If A is a poset satisfying DP1 and DP3, then $l \leq 1$.*

Proof. Suppose to the contrary that there are elements $a, b \in A$ with $l \geq 2$, and pick a pair of minimal rank. Then a, b both cover c, d one rank lower. But then c, d are both covered by a, b and so have an l value at least 2, a contradiction. ∎

We wish to define up and down operators in any differential poset A. But first we need a lemma to make them well defined.

Lemma 5.1.12 *If A is differential, then the nth rank, A_n, is finite for all $n \geq 0$.*

Proof. Induct on n, where the case $n = 0$ is taken care of by DP1. Assume that all ranks up through A_n are finite. Any $a \in A_n$ can cover at most $|A_{n-1}|$ elements, and so can be covered by at most $|A_{n-1}| + 1$ elements by DP2. It follows that
$$|A_{n+1}| \leq |A_n| \left(|A_{n-1}| + 1 \right)$$
forcing A_{n+1} to be finite. ∎

Definition 5.1.13 Let A be a poset where, for any $a \in A$, the sets $\{b \in A : b \prec a\}$ and $\{b \in A : b \succ a\}$ are finite. Consider the vector space $\mathbb{C}A$ of formal linear combinations of elements of A. Define two linear operators, *down*, D, and *up*, U, in $\mathbb{C}A$ by

$$D(\mathbf{a}) = \sum_{b \prec a} \mathbf{b}$$

$$U(\mathbf{a}) = \sum_{b \succ a} \mathbf{b}$$

and linear extension. Note that if A is differential, then by Lemma 5.1.12 these operators are well defined. ∎

The proof of the next result is similar to that of Proposition 5.1.5 and so is left to the reader.

Proposition 5.1.14 *Let A be a graded poset with A_n finite for all $n \geq 0$. Then the following are equivalent.*

    1. A is differential.

    2. $DU - UD = I$. ∎

In fact, differential posets were so named because of the identity $DU - UD = I$. Specifically, if we let D be differentiation and x be the multiplication by x operator acting on the polynomial ring $\mathbb{C}[x]$, then $Dx - xD = I$. The algebra of operators generated by D and x modulo this relation is called the *Weyl algebra*.

We can now generalize Theorem 5.1.8. Again, the demonstration is close to that of the original result, and so the reader will have no trouble filling in the details.

Theorem 5.1.15 ([Stn 88]) *In any differential poset A,*

$$\sum_{a \in A_n} (f^a)^2 = n!,$$

where f^a is the number of saturated $\emptyset$–a chains. ∎

5.2 Growths and Local Rules

It would be nice to give a bijective proof of Theorem 5.1.15 as the Robinson-Schensted algorithm does for Theorem 5.1.8. This was done as part of a theory developed by Fomin [Fom 86, Fom 95], which we now present. For other expositions, see Roby's thesis [Rob 91] or Stanley's text [Stn 99, pp. 327–330].

The basic definition is as follows.

Definition 5.2.1 ([Fom 95]) *Let A, B be posets. Then a growth is a function $g : A \to B$ that preserves the relation "covered by or equal to," which means*

$$a \preceq b \Rightarrow g(a) \preceq g(b). \quad ∎$$

Note that such a function is *order preserving*, i.e., $a \leq b$ implies $g(a) \leq g(b)$, but need not be injective. The term "growth" comes from biology: If $A = C_n$, then we can think of A's elements as units of time and the value $g(i)$ as the state of the system studied at time i. So at time $i + 1$ the system either stays the same or grows to a subsequent level.

Given posets A, B, define their *product*, $A \times B$, as having elements (a, b) with $a \in A$, $b \in B$ ordered by

$$(a, b) \leq (a', b') \text{ in } A \times B \text{ if } a \leq a' \text{ in } A \text{ and } b \leq b' \text{ in } B.$$

We will be particularly concerned with growth functions $g : C_n^2 \to A$ where $C_n^2 = C_n \times C_n$ and A is differential. We will represent the elements of C_n^2 as vertices (i, j), $0 \leq i, j \leq n$, in the first quadrant of the plane and the covering relations as lines (rather than arrows) connecting the vertices. The squares thus formed will also be coordinatized, with square (i, j) being the one whose northeast vertex is (i, j). A permutation $\pi = x_1 x_2 \ldots x_n$ is then represented by putting an X in square (i, x_i), $1 \leq i \leq n$. The next figure illustrates the setup for the permutation

$$\pi = 4\,2\,3\,6\,5\,1\,7.$$

The reader should compare the following construction with Viennot's in Section 3.6.

We now define a growth $g_\pi : C_n^2 \to Y$ that will depend on π. We start by letting

$$g_\pi(i, j) = \emptyset \text{ if } i = 0 \text{ or } j = 0.$$

Consider the square s with coordinates (i, j), labeled as in the diagram

Suppose, by induction on $i+j$, that we have defined $g_\pi(i-1,j-1)$, $g_\pi(i,j-1)$, and $g_\pi(i-1,j)$, which we denote by λ, μ, and ν, respectively. We then define $g_\pi(i,j)$, denoted by ρ, using the following *local rules*

LR1 If $\mu \neq \nu$, then let $\rho = \mu \cup \nu$.

LR2 If $\lambda \prec \mu = \nu$, then μ is obtained from λ by adding 1 to λ_i for some i. Let ρ be obtained from μ by adding 1 to μ_{i+1}.

LR3 If $\lambda = \mu = \nu$, then let

$$\rho = \begin{cases} \lambda & \text{if } s \text{ does not contain an } X, \\ \lambda \text{ with 1 added to } \lambda_1 & \text{if } s \text{ does contain an } X. \end{cases}$$

Applying these rules to our example, we obtain the following diagram, where for readability, parentheses and commas have been suppressed in the partitions:

∅	1	11	21	31	32	321	421 (X)
∅	1	11	21	31 (X)	32	321	321
∅	1	11	21	21	31 (X)	311	311
∅	1 (X)	11	21	21	21	211	211
∅	∅	1	2 (X)	2	2	21	21
∅	∅	1 (X)	1	1	1	11	11
∅	∅	∅	∅	∅	∅	1 (X)	1
∅	∅	∅	∅	∅	∅	∅	∅

Our first order of business is to make sure that g_π is well-defined.

Lemma 5.2.2 *Rules LR1–3 construct a well-defined growth $g_\pi : C_n^2 \to Y$ in that ρ is a partition and satisfies $\rho \succeq \mu, \nu$.*

Proof. The only case where it is not immediate that ρ is a partition is LR2. But by the way μ is obtained from λ we have $\mu_i > \mu_{i+1}$. So adding 1 to μ_{i+1} will not disturb the fact that the parts are weakly decreasing.

It is clear that ρ covers or equals μ, ν except in LR1. Since $\lambda \preceq \mu, \nu$ by induction and $\mu \neq \nu$, there are three possibilities, namely $\lambda = \mu \prec \nu$, $\lambda = \nu \prec \mu$, or $\lambda \prec \mu, \nu$. It is easy to check that in each case, remembering the proof of (2) in Proposition 5.1.3 for the third one, that $\rho \succeq \mu, \nu$. ∎

The following lemma will also be useful. Its proof is a straightforward application of induction and LR1–3, and so is left to the reader

Lemma 5.2.3 *The growth g_π has the following properties.*

1. *$g_\pi(i, j - 1) \prec g_\pi(i, j)$ if and only if there is an X in square (k, j) for some $k \leq i$.*

2. *$g_\pi(i - 1, j) \prec g_\pi(i, j)$ if and only if there is an X in square (i, l) for some $l \leq j$.* ∎

By this lemma, we have a saturated chain of partitions up the eastern boundary of C_n^2,

$$\emptyset = g_\pi(n, 0) \prec g_\pi(n, 1) \prec \cdots \prec g_\pi(n, n),$$

corresponding to a standard tableau that we will denote by P_π. Similarly, along the northern boundary we have Q_π defined by

$$\emptyset = g_\pi(0, n) \prec g_\pi(1, n) \prec \cdots \prec g_\pi(n, n).$$

In our running example,

$$P_\pi = \begin{matrix} 1\ 3\ 5\ 7 \\ 2\ 6 \\ 4 \end{matrix} \quad \text{and} \quad Q_\pi = \begin{matrix} 1\ 3\ 4\ 7 \\ 2\ 5 \\ 6 \end{matrix}.$$

Comparing this with the tableaux obtained on page 94 (computed by insertion and placement) gives credence to the following theorem.

Theorem 5.2.4 *The map*

$$\pi \to (P_\pi, Q_\pi)$$

is the Robinson-Schensted correspondence.

Proof. We will prove that $P_\pi = P(\pi)$ since the equation $Q_\pi = Q(\pi)$ is derived using similar reasoning.

First we will associate with each (i, j) in C_n^2 a permutation $\pi_{i,j}$ and a tableau $P_{i,j}$. Note that in $P_{i,j}$, the subscripts refer to the element of C_n^2 and not the entry of the tableau in row i and column j as is our normal convention.

Since this notation will be used only in the current proof, no confusion will result. Define $\pi_{i,j}$ as the lower line of the partial permutation corresponding to those X's southwest of (i,j) in the diagram for g_π. In our usual example,

$$\pi_{6,3} = 2\ 3\ 1.$$

We also have a sequence of shapes

$$\lambda_{i,0} \preceq \lambda_{i,1} \preceq \cdots \preceq \lambda_{i,j},$$

where $\lambda_{i,l} = g_\pi(i,l)$. So $|\lambda_{i,l}/\lambda_{i,l-1}| = 0$ or 1 for $1 \le l \le j$. In the latter case, label the box of $\lambda_{i,l}/\lambda_{i,l-1}$ with l to obtain the tableau $P_{i,j}$. These tableaux are illustrated for our example in the following figure:

∅	4	2 4	23 4	236 4	235 46	135 26 4	1357 26 4 X^4
∅	4	2 4	23 4	236 4 X	235 46	135 26 4	135 26 4
∅	4	2 4	23 4	23 4	235 4 X	135 2 4	135 2 4
∅	4 X	2 4	23 4	23 4	23 4	13 2 4	13 2 4
∅	∅	2	23 X	23	23	13 2	13 2
∅	∅	2 X	2	2	2	1 2	1 2
∅	∅	∅	∅	∅	∅ X	1	1
∅	∅	∅	∅	∅	∅	∅	∅

It is clear from the definitions that $\pi_{n,n} = \pi$ and $P_{n,n} = P_\pi$. So it suffices to prove that $P_{i,j} = P(\pi_{i,j})$ for all $i,j \le n$. We will do this by induction on $i + j$. It is obvious if $i = 0$ or $j = 0$. The proof breaks down into three cases depending on the coordinates (i,l) of the X in row i, and (k,j) of the X in column j.

Case 1: $l > j$. (The case $k > i$ is similar.) By Lemma 5.2.3, we have $\lambda_{i,j-1} = \lambda_{i,j}$, so $P_{i,j-1} = P_{i,j}$. Also, $\pi_{i,j-1} = \pi_{i,j}$. Since $P_{i,j-1} = P(\pi_{i,j-1})$ by induction, we are done with this case.

Case 2: $l = j$. (Note that this forces $k = i$.) By Lemma 5.2.3 again, $\lambda_{i-1,j-1} = \lambda_{i,j-1} = \lambda_{i-1,j}$. So the second option in LR3 applies, and $P_{i,j}$ is $P_{i,j-1}$ with j placed at the end of the first row. Furthermore, $\pi_{i,j}$ is the concatenation $\pi_{i,j-1}j$, where $j > \max \pi_{i,j-1}$. So $P(\pi_{i,j})$ is also $P(\pi_{i,j-1})$ with j placed at the end of the first row, which finishes this case.

Case 3: $l < j$. (We can assume $k < i$, since the other possibilities are taken care of by previous cases.) Using Lemma 5.2.3 one more time, we have $\lambda_{i-1,j-1} \prec \lambda_{i,j-1}, \lambda_{i-1,j}$. Let c and d denote the cells of the skew partitions $\lambda_{i,j-1}/\lambda_{i-1,j-1}$ and $\lambda_{i-1,j}/\lambda_{i-1,j-1}$, respectively. If $c \neq d$, then rule LR1 is the relevant one and $P_{i,j}$ is $P_{i,j-1}$ with j placed in cell d. If $c = d$, then LR2 implies that $P_{i,j}$ is $P_{i,j-1}$ with j placed at the end of the next row below that of d.

As far as the permutations are concerned, there are sequences σ, τ such that

$$
\begin{aligned}
\pi_{i-1,j-1} &= \sigma\tau, \\
\pi_{i,j-1} &= \sigma\tau l, \\
\pi_{i-1,j} &= \sigma j\tau. \\
\pi_{i,j} &= \sigma j\tau l,
\end{aligned}
$$

where, as before, $j > \max \pi_{i,j-1}$. So the insertion paths for $P(\pi_{i,j-1})$ and $P(\pi_{i,j})$ are the same except that when a path of the former ends at the row currently occupied by j in the latter, then j is bumped to the next row. Now, by induction, j ends up in cell d of $P(\pi_{i-1,j})$, and the insertion path of l into $P(\pi_{i-1,j-1})$ ends in cell c of $P(\pi_{i,j-1})$. Thus if $c \neq d$, then j is not displaced by the insertion of l, and $P(\pi_{i,j})$ is $P(\pi_{i,j-1})$ with j placed in cell d. And if $c = d$, then j is displaced by r_l into the row below d. For both options, the description of $P(\pi_{i,j})$ matches that of $P_{i,j}$, so this completes the proof of the theorem. ∎

We are now in a position to define growths $g_\pi : C_n^2 \to A$ where A is any differential poset. The initial conditions should come as no surprise:

$$g_\pi(i,j) = \hat{0} \text{ if } i = 0 \text{ or } j = 0.$$

We label the square s with coordinates (i,j) as follows, where a, b, c are given and d is to be defined:

To state the local rules, note that from condition DP2 we have, for fixed b,

$$|\{x \;:\; x \preceq b\}| = |\{y \;:\; y \succ b\}|.$$

So for each b, pick a bijection

$$\Psi_b : \{x \ : \ x \preceq b\} \to \{y \ : \ y \succ b\}.$$

The growth g_π will depend on the Ψ_b, although our notation will not reflect that fact.

DLR1 If $b \neq c$, then $a = b \prec c$, $a = c \prec b$, or $a \prec b, c$. In the first two cases let $d = \max\{b, c\}$. In the third let d be the unique element covering b, c as guaranteed by DP3 and Lemma 5.1.11.

DLR2 If $a \prec b = c$, then let $d = \Psi_b(a)$.

DLR3 If $a = b = c$, then let

$$d = \begin{cases} a & \text{if } s \text{ does not contain an } X, \\ \Psi_b(a) & \text{if } s \text{ does contain an } X. \end{cases}$$

It is easy to show that these rules are well-defined.

Lemma 5.2.5 *Rules DLR1–3 construct a well-defined growth* $g_\pi : C_n^2 \to A$ *in that d satisfies $d \succeq b, c$.* ∎

Furthermore, Lemma 5.2.3 continues to hold in this more general setting. So, as before, we have saturated chains of partitions across the eastern and northern boundaries of C_n^2, which we will denote by C_π and $\mathcal{D}_\pi$, respectively. So Theorem 5.1.15 is a corollary of the following result.

Theorem 5.2.6 *The map*

$$\pi \to (C_\pi, \mathcal{D}_\pi)$$

is a bijection between S_n and pairs of saturated $\hat{0}$-a chains as a varies over all elements of rank n in the differential poset A.

Proof. Since both C and $\mathcal{D}$ end at $(n, n) \in C_n^2$, it is clear that the corresponding chains end at the same element which must be of rank n.

To show that this is a bijection, we construct an inverse. Given saturated chains

$$\begin{aligned} C : \hat{0} &= c_0 \prec c_1 \prec \cdots \prec c_n \\ \mathcal{D} : \hat{0} &= d_0 \prec d_1 \prec \cdots \prec d_n \end{aligned}$$

where $c_n = d_n$, define

$$g_\pi(n, j) = c_j \text{ and } g_\pi(i, n) = d_i \text{ for } 0 \leq i, j \leq n.$$

Also, let $\Phi_b = \Psi_b^{-1}$. Assuming that b, c, d are given around a square s as in the previous diagram, we define a as follows.

RLD1 If $b \neq c$, then $d = b \succ c$, $d = c \succ b$, or $d \succ b, c$. In the first two cases let $a = \min\{b, c\}$. In the third let a be the unique element covered by b, c.

RLD2 If $d \succ b = c$, then let $a = \Phi_b(d)$.

RLD3 If $b = c = d$, then let $a = d$.

Finally, we recover π as corresponding to those squares where $a = b = c \prec d$. Showing that RLD1–3 give a well-defined growth associated with some permutation π and that this is the inverse procedure to the one given by DLR1–3 is straightforward and left to the reader. ∎

5.3 Groups Acting on Posets

The systematic study of group actions on posets was initiated by Stanley [Stn 82] and has blossomed in recent years [Han 81, C-H-R 86, Sun 94, B-W 95, S-Wa 97, S-We]. In this section and the next we will investigate some of the interesting representations that arise in this context and some of their combinatorial implications.

Definition 5.3.1 Let A be a finite, graded poset of rank n, having miminum $\hat{0}$ and maximum $\hat{1}$. Any $S \subseteq \{1, 2, \ldots, n-1\}$ defines a corresponding *rank-selected subposet*

$$A_S = \{a \in A \ : \ a = \hat{0}, \ a = \hat{1}, \ \text{or} \ \text{rk} \, a \in S\}$$

with partial order $a \leq b$ in A_S if and only if the same inequality holds in A. We will use the notation $S = \{s_1, s_2, \ldots, s_k\}_<$ to mean $s_1 < s_2 < \cdots < s_k$. ∎

For example, if A is the Boolean algebra B_4 and $S = \{1, 3\}$, then the Hasse diagram of A_S is given on the next page. All posets considered in this section will be finite, graded, with a $\hat{0}$ and a $\hat{1}$, and so will admit rank-selections.

In order to define the invariants we wish to study, we will have to enlarge our definition of representation.

Definition 5.3.2 Let group G have irreducible modules $V^{(1)}, V^{(2)}, \ldots, V^{(k)}$. A *virtual representation* of G is a formal sum

$$V = \sum_i m_i V^{(i)}$$

where the multiplicities m_i are in $\mathbb{Z}$, the integers. ∎

If the m_i are nonnegative, then V is an ordinary representation and the sum can be interpreted as being direct.

Now let G be a group of *automorphisms* of the poset A, i.e., a group of bijections $g : A \to A$ such that both g and g^{-1} are order preserving. Then it is easy to see that each $g \in G$ permutes the maximal (with respect to containment) chains of A_S. Let A^S be the associated permutation module for G. Define a virtual representation B^S by

$$B^S = \sum_{T \subseteq S} (-1)^{|S-T|} A^T. \tag{5.6}$$

We should remark that Stanley [Stn 82] defined B^S in terms of G's action on the reduced homology groups of A_S and then derived (5.6) as a corollary. However, we will have no need here for that viewpoint.

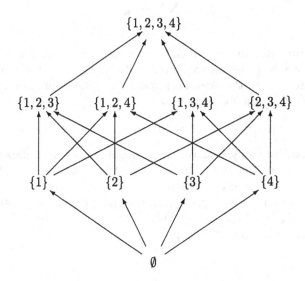

We would also like to express A^S in term of the B^T. This will be done with the aid of the next result.

Theorem 5.3.3 *Let V be a vector space and suppose we have two functions $f, g : B_n \to V$. Then the following are equivalent:*

$$(1) \quad f(S) \;=\; \sum_{T \subseteq S} g(T) \text{ for all } S \in B_n,$$

$$(2) \quad g(S) \;=\; \sum_{T \subseteq S} (-1)^{|S-T|} f(T) \text{ for all } S \in B_n.$$

Proof. We will prove that (1) implies (2), as the converse is similar. Now,

$$\sum_{T \subseteq S} (-1)^{|S-T|} f(T) \;=\; \sum_{T \subseteq S} (-1)^{|S-T|} \sum_{U \subseteq T} g(U)$$
$$= \sum_{U \subseteq S} (-1)^{|S-U|} g(U) \sum_{U \subseteq T \subseteq S} (-1)^{|T-U|}.$$

But from Exercise 1e in Chapter 4 we see that the inner sum is 0 unless $U = S$. So the outer sum collapses to $g(S)$. ∎

Theorem 5.3.3 is equivalent to the Principle of Inclusion-Exclusion (see Exercise 17) which in turn is the special case of Möbius inversion (see Exercise 18) where the underlying poset is the Boolean algebra [Sag 99]. As an immediate corollary of this Theorem and equation (5.6), we have the following result.

Corollary 5.3.4 *For all $S \subseteq \{1, 2, \ldots, n-1\}$ we have*

$$A^S = \sum_{T \subseteq S} B^T. \quad \blacksquare$$

If the B^S turn out to be ordinary representations, then something interesting is probably going on combinatorially. This will be illustrated by turning our attention to the Boolean algebra B_n. The symmetric group $\mathcal{S}_n$ acts on the $S \in B_n$ in the natural way by

$$\pi\{n_1, n_2, \ldots, n_k\} = \{\pi n_1, \pi n_2, \ldots, \pi n_k\}, \tag{5.7}$$

and it is not hard to verify that this is an action, called an *induced action*, which is an automorphism of B_n. In this case, if $S = \{n_1, n_2, \ldots, n_k\}_<$, then A^S is equivalent to the representation afforded by the permutation module M^μ, where $\mu = (n_1, n_2 - n_1, n_3 - n_2, \ldots, n - n_k)$. To see this, note that if $R_1 \uplus R_2 \uplus \cdots \uplus R_{k+1} = \{1, 2, \ldots, n\}$ with $|R_i| = n_i - n_{i-1}$ for $1 \leq i \leq k+1$ ($n_0 = 0$ and $n_{k+1} = n$), then the map

$$C : \emptyset < R_1 < R_1 \uplus R_2 < \cdots < \{1, 2, \ldots, n\} \mapsto \{t\} = \frac{\boxed{\begin{array}{c} R_1 \\ \hline R_2 \\ \hline \vdots \\ \hline R_{k+1} \end{array}}}{} \tag{5.8}$$

extends linearly to an $\mathcal{S}_n$-module isomorphism.

To describe B^S, we need one more definition.

Definition 5.3.5 Let P be a standard tableau. Then $i \in P$ is a *descent* if $i+1$ is in a lower row of P. The *descent set of P* is

$$\mathrm{Des}\, P = \{i \; : \; i \text{ is a descent of } P\}. \quad \blacksquare$$

For example,

$$\text{if} \quad P = \begin{array}{cccc} 1 & 2 & 3 & 5 \\ 4 & 6 & 8 & \\ 7 & & & \end{array}, \quad \text{then} \quad \mathrm{Des}\, P = \{3, 5, 6\}.$$

Counting descents is intimately connected with Kostka numbers.

Proposition 5.3.6 *Let $\lambda \vdash n$, $S = \{n_1, n_2, \ldots, n_k\}_< \subseteq \{1, 2, \ldots, n-1\}$, and $\mu = (n_1, n_2 - n_1, \ldots, n - n_k)$. Then*

$$|\{P \; : \; P \text{ a standard } \lambda\text{-tableau and } \mathrm{Des}\, P \subseteq S\}| = K_{\lambda\mu}.$$

Proof. We will give a bijection between the two sets involved, where $K_{\lambda\mu}$ counts semistandard tableaux T of shape λ and content μ. Given P, replace the numbers $1, 2, \ldots, n_1$ with ones, then $n_1 + 1, n_1 + 2, \ldots, n_2$ with twos, and so on. The next figure illustrates the map when $\lambda = (4, 3, 1)$, $S = \{3, 5, 7\}$, and so $\mu = (3, 2, 2, 1)$. Below each T is the corresponding P, and below that, Des P is given:

$$
T: \quad
\begin{matrix} 1\,2\,3\,5 \\ 4\,6\,7 \\ 8 \end{matrix}
\,,\quad
\begin{matrix} 1\,2\,3\,5 \\ 4\,7\,8 \\ 6 \end{matrix}
\,,\quad
\begin{matrix} 1\,2\,3\,7 \\ 4\,5\,6 \\ 8 \end{matrix}
\,,\quad
\begin{matrix} 1\,2\,3\,7 \\ 4\,5\,8 \\ 6 \end{matrix}
\,,\quad
\begin{matrix} 1\,2\,3\,8 \\ 4\,5\,7 \\ 6 \end{matrix}
\;;
$$

$$
P: \quad
\begin{matrix} 1\,1\,1\,2 \\ 2\,3\,3 \\ 4 \end{matrix}
\,,\quad
\begin{matrix} 1\,1\,1\,2 \\ 2\,3\,4 \\ 3 \end{matrix}
\,,\quad
\begin{matrix} 1\,1\,1\,3 \\ 2\,2\,3 \\ 4 \end{matrix}
\,,\quad
\begin{matrix} 1\,1\,1\,3 \\ 2\,2\,4 \\ 3 \end{matrix}
\,,\quad
\begin{matrix} 1\,1\,1\,4 \\ 2\,2\,3 \\ 3 \end{matrix}
\;;
$$

$$
\text{Des}\,P \quad \{3,5,7\}, \quad \{3,5\}, \quad \{3,7\}, \quad \{3,5,7\}, \quad \{3,5\}.
$$

The resulting T is clearly a λ-tableau of content μ, so we need only check that it is semistandard. Since the rows of P increase and the replacement map is weakly increasing, the rows of T must weakly increase. Similarly, the columns of T weakly increase. To show that they actually increase strictly, suppose $T_{i,j} = T_{i+1,j} = l$. Then at least one of $n_{l-1} + 1, n_{l-1} + 2, \ldots, n_l - 1$ is a descent in P, contradicting Des $P \subseteq S$.

To show that this map is a bijection, we must construct an inverse. This is a simple matter and left as an exercise. ∎

We can now describe the decomposition of B^S into irreducibles for the Boolean algebra and show that in this case it is a genuine (as opposed to virtual) representation.

Theorem 5.3.7 ([Stn 82]) *For the action of $\mathcal{S}_n$ on B_n, decompose*

$$
B^S \cong \sum_\lambda b^S(\lambda) S^\lambda.
$$

Then the muliplicities are nonnegative integers, since

$$
b^S(\lambda) = |\{P \; : \; P \text{ a standard } \lambda\text{-tableau and Des } P = S\}|.
$$

Proof. Decompose A^S as

$$
A^S \cong \sum_\lambda a^S(\lambda) S^\lambda.
$$

Then Corollary 5.3.4 implies

$$
a^S(\lambda) = \sum_{T \subseteq S} b^T(\lambda) \quad \text{for all} \quad S \subseteq \{1, 2, \ldots, n-1\}. \tag{5.9}
$$

Define $\bar{a}^S(\lambda)$ (respectively, $\bar{b}^S(\lambda)$) to be the number of standard λ-tableaux P with Des $P \subseteq S$ (respectively, Des $P = S$). Then directly from the definitions

$$\bar{a}^S(\lambda) = \sum_{T \subseteq S} \bar{b}^T(\lambda) \quad \text{for all} \quad S \subseteq \{1, 2, \ldots, n-1\}. \tag{5.10}$$

Now, from Proposition 5.3.6 and the fact that A^S is equivalent to M^μ we have $a^S(\lambda) = K_{\lambda\mu} = \bar{a}^S(\lambda)$ for all S. But then inverting (5.9) and (5.10) using Theorem 5.3.3 shows that $b^S(\lambda) = \bar{b}^S(\lambda)$ for all S, as desired. ∎

5.4 Unimodality

Unimodality of a sequence of real numbers means that it rises to some maximum and then descends. It is the next most complicated behavior after being weakly increasing or weakly decreasing. Unimodal sequences abound in algebra, combinatorics, and geometry. See Stanley's article [Stn 89] for a survey. In this section we will show how group actions can be used to prove unimodality.

Definition 5.4.1 A sequence $(a_k)_{k \geq 0} = a_0, a_1, a_2, \ldots$ of real numbers is *unimodal* if, for some index m,

$$a_0 \leq a_1 \leq \cdots \leq a_m \geq a_{m+1} \geq \cdots.$$

If the sequence is finite with n terms, then it is *symmetric* if

$$a_k = a_{n-k} \quad \text{for} \quad 0 \leq k \leq n.$$

A poset A with finite rank sets is *symmetric* or *unimodal* whenever the sequence $(|A_k|)_{k \geq 0}$ is. ∎

Our first example of a symmetric, unimodal sequence comes from the binomial coefficients.

Proposition 5.4.2 *For any n, the sequence*

$$\binom{n}{0}, \binom{n}{1}, \ldots, \binom{n}{n}$$

is symmetric and unimodal. It follows that the same is true of the Boolean algebra B_n.

Proof. For symmetry we have, from the factorial formula in Exercise 1b of Chapter 4,

$$\binom{n}{n-k} = \frac{n!}{(n-k)!(n-(n-k))!} = \frac{n!}{(n-k)!(k)!} = \binom{n}{k}.$$

Since the sequence is symmetric, we can prove unimodality by showing that it increases up to its midpoint. Using the same exercise, we see that for $k \leq n/2$:

$$\frac{\binom{n}{k-1}}{\binom{n}{k}} = \frac{n!k!(n-k)!}{n!(k-1)!(n-k+1)!} = \frac{k}{n-k+1} < 1. \quad \blacksquare$$

It will also be useful to talk about unimodal sequences of representations.

Definition 5.4.3 Let V, W be G-modules whose expansions in terms of the G's irreducibles are $V \cong \sum_{i \geq 1} m_i V^{(i)}$ and $W \cong \sum_{i \geq 1} n_i V^{(i)}$. Then we write $V \leq W$ if $m_i \leq n_i$ for all $i \geq 1$. $\blacksquare$

Using this partial order, we define a *unimodal sequence of modules* in the obvious way.

To get permutation groups into the act, consider a finite graded poset, A. Let U_k be the up operator of Definition 5.1.13 restricted to $\mathbb{C}A_k$. Also, let $X = X_k$ be the matrix of U_k in the bases $\{\mathbf{a} : a \in A_k\}$ and $\{\mathbf{b} : b \in A_{k+1}\}$, so

$$X_{a,b} = \begin{cases} 1 & \text{if } a \prec b, \\ 0 & \text{else.} \end{cases}$$

Finally, let $\operatorname{rk} X$ denote the (linear-algebraic) rank of the matrix X.

Lemma 5.4.4 *Let G be a group of automorphisms of a finite, graded poset A.*

1. *If $\operatorname{rk} X = |A_k|$, then $\mathbb{C}A_k \leq \mathbb{C}A_{k+1}$.*

2. *If $\operatorname{rk} X = |A_{k+1}|$, then $\mathbb{C}A_k \geq \mathbb{C}A_{k+1}$.*

Proof. We will prove the first statement, as the second is similar. We claim that U_k is actually a G-homomorphism. It suffices to show that $U_k(g\mathbf{a}) = gU_k(\mathbf{a})$ for $g \in G$ and $a \in A_k$. In other words, it suffices to prove that

$$\{b : b \succ ga\} = \{gc : c \succ a\}.$$

But if $b \succ ga$, then letting $c = g^{-1}b$ we have $gc = b \succ ga$, so $c \succ a$. And if $c \succ a$, then $b = gc \succ ga$, so the two sets are equal.

Since $\operatorname{rk} X = |A_k|$, we have that U_k is an injective G-homomorphism, and so $\mathbb{C}A_k$ is isomorphic to a submodule of $\mathbb{C}A_{k+1}$. Since G is completely irreducible, the inequality follows. $\blacksquare$

Definition 5.4.5 A finite, graded poset A of rank n is *ample* if

$$\operatorname{rk} X_k = \min\{|A_k|, |A_{k+1}|\} \quad \text{for} \quad 0 \leq k < n. \quad \blacksquare$$

This is the definition needed to connect unimodality of posets and modules. We will also be able to tie in the concept of orbit as defined in Exercise 2 of Chapter 1. If G acts on a finite set S, then the *orbit* of $s \in S$ is

$$\mathcal{O}_s = \{gs \: : \: g \in G\}.$$

Also, let

$$S/G = \{\mathcal{O}_s \: : \: s \in S\}.$$

Theorem 5.4.6 ([Stn 82]) *Let A be a finite, graded poset of rank n. Let G be a group of automorphisms of A and V be an irreducible G-module. If A is unimodal and ample, then the following sequences are unimodal.*

(1) $\mathbb{C}A_0, \mathbb{C}A_1, \mathbb{C}A_2, \ldots, \mathbb{C}A_n$.

(2) $m_0(V), m_1(V), m_2(V), \ldots, m_n(V)$, *where $m_k(V)$ is the multiplicity of V in $\mathbb{C}A_k$.*

(3) $|A_0/G|, |A_1/G|, |A_2/G|, \ldots, |A_n/G|$.

Proof. The fact that the first sequence is unimodal follows immediately from the definition of ample and Lemma 5.4.4. This implies that the second sequence is as well by definition of the partial order on G-modules. Finally, by Exercise 5b in Chapter 2, (3) is the special case of (2) where one takes V to be the trivial module. ∎

In order to apply this theorem to the Boolean algebra, we will need the following lemma. The proof we give is based on Kantor's [Kan 72]. Another is indicated in Exercise 25.

Proposition 5.4.7 *The Boolean algebra B_n is ample.*

Proof. By Proposition 5.4.2 and the fact that X_{n-k-1} is the transpose of X_k, it suffices to show that $\operatorname{rk} X_k = \binom{n}{k}$ for $k < n/2$. Let X_S denote the row of $X = X_k$ indexed by the set S. Suppose, towards a contradiction, that there is a dependence relation

$$X_{\overline{S}} = \sum_{S \neq \overline{S}} a_S X_S \tag{5.11}$$

for certain real numbers a_S, not all zero. Let G be the subgroup of $\mathcal{S}_n$ that fixes $\overline{S}$ as a set, i.e., $G = \mathcal{S}_{\overline{S}} \times \mathcal{S}_{\overline{S}^c}$, where $\overline{S}^c$ is the complement of $\overline{S}$ in $\{1, 2, \ldots, n\}$. Then G acts on the kth rank

$$B_{n,k} \stackrel{\text{def}}{=} (B_n)_k,$$

and if $g \in G$, then $g\overline{S} = \overline{S}$ and $X_{gS,gT} = X_{S,T}$ for all $S \in B_{n,k}, T \in B_{n,k+1}$. Also, g permutes the $S \neq \overline{S}$ among themselves. So, using (5.11), we obtain

$$X_{\overline{S},T} = X_{g\overline{S},gT} = X_{\overline{S},gT} = \sum_{S \neq \overline{S}} a_S X_{S,gT} = \sum_{S \neq \overline{S}} a_{gS} X_{gS,gT} = \sum_{S \neq \overline{S}} a_{gS} X_{S,T}.$$

It follows that $X_{\overline{S}} = \sum_{S \neq \overline{S}} a_{gS} X_S$, and summing over all $g \in G$ gives

$$|G|X_{\overline{S}} = \sum_{g \in G} \sum_{S \neq \overline{S}} a_{gS} X_S = \sum_{S \neq \overline{S}} X_S \sum_{g \in G} a_{gS}. \qquad (5.12)$$

By Exercise 5a in Chapter 2, the action of G partitions $B_{n,k}$ into orbits $\mathcal{O}_i$. Directly from the definitions we see that we can write

$$\mathcal{O}_i = \{S \in B_{n,k} \; : \; |S \cap \overline{S}| = i\}, \quad 0 \leq i \leq k.$$

So $\mathcal{O}_k = \{\overline{S}\}$ and $\uplus_{i=0}^{k-1} \mathcal{O}_i$ is the set of all $S \neq \overline{S}$, $S \in B_{n,k}$. Because of the bijection from Exercise 2b in Chapter 1, if $S \in \mathcal{O}_i$, then as g runs over G, the sets $T = gS$ will run over $\mathcal{O}_i$ with each set being repeated $|G|/|\mathcal{O}_i|$ times. So (5.12) can be rewritten

$$
\begin{aligned}
X_{\overline{S}} &= \frac{1}{|G|} \sum_{i=0}^{k-1} \sum_{S \in \mathcal{O}_i} X_S \sum_{g \in G} a_{gS} \\
&= \sum_{i=0}^{k-1} \sum_{S \in \mathcal{O}_i} X_S \frac{1}{|\mathcal{O}_i|} \sum_{T \in \mathcal{O}_i} a_T \\
&= \sum_{i=0}^{k-1} \frac{1}{|\mathcal{O}_i|} \sum_{T \in \mathcal{O}_i} a_T \sum_{S \in \mathcal{O}_i} X_S \\
&= \sum_{i=0}^{k-1} b_i \sum_{S \in \mathcal{O}_i} X_S
\end{aligned}
$$

for certain real numbers b_i, not all zero (since the sum is a nonzero vector). Now, for $0 \leq j < k$, choose $T_j \in B_{n,k+1}$ with $|T_j \cap \overline{S}| = j$, which can be done since $(k+1)+k = 2k+1 \leq n$. But then $X_{\overline{S},T_j} = 0$. So the previous equation for $X_{\overline{S}}$ gives

$$0 = \sum_{i=0}^{k-1} b_i \sum_{S \in \mathcal{O}_i} X_{S,T_j}, \qquad (5.13)$$

which can be viewed as a system of k equations in k unknowns, namely the b_i. Now, if $i > j$ then $X_{S,T_j} = 0$ for all $S \in \mathcal{O}_i$. So this system is actually triangular. Furthermore, if $i = j$ then there is an $S \in \mathcal{O}_i$ with $X_{S,T_j} = 1$, so that the system has nonzero coefficients along the diagonal. It follows that $b_i = 0$ for all i, the desired contradiction. $\blacksquare$

Corollary 5.4.8 Let $G \leq \mathcal{S}_n$ act on B_n and let V be an irreducible G-module. Then, keeping the notation in the statement and proof of the previous proposition, the following sequences are symmetric and unimodal.

(1) $m_0(V), m_1(V), m_2(V), \ldots, m_n(V)$, where $m_k(V)$ is the multiplicity of V in $\mathbb{C}B_{n,k}$.

(2) $|B_{n,0}/G|, |B_{n,1}/G|, |B_{n,2}/G|, \ldots, |B_{n,n}/G|$.

Proof. The fact that the sequences are unimodal follows from Theorem 5.4.6, and Propositions 5.4.2 and 5.4.7. For symmetry, note that the map $f : B_{n,k} \to B_{n,n-k}$ sending S to its complement induces a G-module isomorphism. ■

As an application of this corollary, fix nonnegative integers k, l and consider the partition (l^k), which is just a $k \times l$ rectangle. There is a corresponding lower order ideal in Young's lattice

$$Y_{k,l} = \{\lambda \in Y \ : \ \lambda \leq (l^k)\}$$

consisting of all partitions fitting inside the rectangle. We will show that this poset is symmetric and unimodal. But first we need an appropriate group action.

Definition 5.4.9 Let G and H be permutation groups acting on sets S and T, respectively. The *wreath product*, $G \wr H$, acts on $S \times T$ as follows. The elements of $G \wr H$ are all pairs of the form (g, h), where $h \in H$ and $g \in G^{|T|}$, so $g = (g_t \ : \ g_t \in G)$. The action is

$$(g, h)(s, t) = (g_t s, ht). \tag{5.14}$$

It is easy to verify that $G \wr H$ is a group and that this is actually a group action. ■

To illustrate, if $G = \mathcal{S}_3$ and $H = \mathcal{S}_2$ acting on $\{1, 2, 3\}$ and $\{1, 2\}$, respectively, then a typical element of $G \wr H$ is

$$(g, h) = (g, (1, 2)), \quad \text{where} \quad g = (g_1, g_2) = ((1, 2, 3), (1, 2, 3)^2).$$

So an example of the action would be

$$(g, h)(3, 2) = (g_2(3), h(2)) = (2, 1).$$

Now, let
$$p_{k,l}(i) = \{\lambda \vdash i \ : \ \lambda \subseteq (l^k)\}.$$

Corollary 5.4.10 *For fixed k, l, the sequence*

$$p_{k,l}(0), p_{k,l}(1), p_{k,l}(2), \ldots, p_{k,l}(kl)$$

is symmetric and unimodal. So the poset $Y_{k,l}$ is as well.

Proof. Identify the cells of (l^k) with the elements of $\{1, 2, \ldots, k\} \times \{1, 2, \ldots, l\}$ in the usual way. Then $\mathcal{S}_k \wr \mathcal{S}_l$ has an induced action on subsets of (l^k), which are partially ordered as in B_{kl} (by containment). There is a unique partition in each orbit, and so the result follows from Corollary 5.4.8, part (2). ■

5.5 Chromatic Symmetric Functions

In this section we will introduce the reader to combinatorial graphs and their proper colorings. It turns out that one can construct a generating function for such colorings that is symmetric and so can be studied in the light of Chapter 4.

Definition 5.5.1 A *graph*, Γ, consists of a finite set of vertices $V = V(\Gamma)$ and a set $E = E(\Gamma)$ of edges, which are unordered pairs of vertices. An edge connecting vertices u and v will be denoted by uv. In this case we say that u and v are *neighbors*. ∎

As an example, the graph in the following figure has

$$V = \{v_1,\, v_2,\, v_3,\, v_4\} \quad \text{and} \quad E = \{e_1,\, e_2,\, e_3,\, e_4\} = \{v_1v_2,\, v_1v_3,\, v_2v_3,\, v_3v_4\},$$

so v_1 is is a neighbor of v_2 and v_3 but not of v_4:

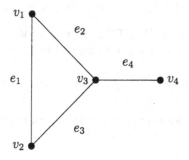

Definition 5.5.2 A *coloring* of a graph Γ from a color set C is a function $\kappa : V \to C$. The coloring is *proper* if it satifies

$$uv \in E \Rightarrow \kappa(u) \neq \kappa(v).\ \blacksquare$$

So a coloring just assigns a color to each vertex and, it is proper if no two vertices of the same color are connected by an edge. One coloring of our graph from the color set $C = \{1, 2, 3\}$ is

$$\kappa(v_1) = \kappa(v_2) = \kappa(v_4) = 1 \quad \text{and} \quad \kappa(v_3) = 2.$$

It is not proper, since both vertices of edge v_1v_2 have the same color. The proper colorings of Γ are exactly those where v_1, v_2, v_3 all have distinct colors and v_4 has a color different from that of v_3. Perhaps the most famous (and controversial because of its computer proof by Appel and Haken) theorem about proper colorings is the Four Color Theorem. It is equivalent to the fact that one can always color a map with four colors so that adjacent countries are colored differently.

Theorem 5.5.3 (Four Color Theorem [A-H 76]) *If graph Γ is planar (i.e., can be drawn in the plane without edge crossings), then there is a proper coloring $\kappa : V \to \{1, 2, 3, 4\}$.* ∎

Definition 5.5.4 The *chromatic polynomial of* Γ is defined to be

$$P_\Gamma = P_\Gamma(n) = \text{ the number of proper } \kappa : V \to \{1, 2, \ldots, n\}. \quad \blacksquare$$

The chromatic polynomial provides an interesting tool for studying proper colorings. Note that Theorem 5.5.3 can be restated: If Γ is planar, then $P_\Gamma(4) \geq 1$. To compute $P_\Gamma(n)$ in our running example, if we color the verticies in order of increasing subscript, then there are n ways to color v_1. This leaves $n - 1$ choices for v_2, since $v_1 v_2 \in E$, and $n - 2$ for v_3. Finally, there are $n - 1$ colors available for v_4, namely any except $\kappa(v_3)$. So in this case $P_\Gamma(n) = n(n - 1)^2(n - 2)$ is a polynomial in n. This is always true, as suggested by the terminology, but is not obvious. The interested reader will find two different proofs in Exercises 28 and 29.

In order to bring symmetric functions into play, one needs to generalize the chromatic polynomial. This motivates the following definition of Stanley.

Definition 5.5.5 ([Stn 95]) If Γ has vertex set $V = \{v_1, \ldots, v_d\}$, then the *chromatic symmetric function of* Γ is defined to be

$$X_\Gamma = X_\Gamma(\mathbf{x}) = \sum_{\kappa : V \to \mathbb{P}} x_{\kappa(v_1)} x_{\kappa(v_2)} \cdots x_{\kappa(v_d)},$$

where the sum is over all proper colorings κ with colors from $\mathbb{P}$, the positive integers. ∎

It will be convenient to let $\mathbf{x}^\kappa = x_{\kappa(v_1)} x_{\kappa(v_2)} \cdots x_{\kappa(v_d)}$. If we return to our example, we see that given any four colors we can assign them to V in $4! = 24$ ways. If we have three colors with one specified as being used twice, then it is not hard to see that there are 4 ways to color Γ. Since these are the only possibilities,

$$
\begin{aligned}
X_\Gamma(\mathbf{x}) &= 24x_1 x_2 x_3 x_4 + 24x_1 x_2 x_3 x_5 + \cdots + 4x_1^2 x_2 x_3 + 4x_1 x_2^2 x_3 + \cdots \\
&= 24m_{1^4} + 4m_{2,1^2}.
\end{aligned}
$$

We next collect a few elementary properties of X_Γ, including its connection to the chromatic polynomial.

Proposition 5.5.6 ([Stn 95]) *Let Γ be any graph with vertex set V.*

1. *X_Γ is homogeneous of degree $d = |V|$.*

2. *X_Γ is a symmetric function.*

3. *If we set $x_1 = \cdots = x_n = 1$ and $x_i = 0$ for $i > n$, written $\mathbf{x} = 1^n$, then*

$$X_\Gamma(1^n) = P_\Gamma(n).$$

Proof. 1. Every monomial in X_Γ has a factor for each vertex.

2. Any permutation of the colors of a proper coloring gives another proper coloring. This means that permuting the subscripts of X_Γ leaves the function invariant. And since it is homogeneous of degree d, it is a *finite* sum of monomial symmetric functions.

3. With the given specialization, each monomial of X_Γ becomes either 1 or 0. The surviving monomials are exactly those that use only the first n colors. So their sum is $P_\Gamma(n)$ by definition of the chromatic polynomial as the number of such colorings. ∎

Since $X_\Gamma \in \Lambda$, we can consider its expansion in the various bases introduced in Chapter 4. In order to do so, we must first talk about partitions of a *set* S.

Definition 5.5.7 A *set partition* $\beta = B_1/B_2/\ldots/B_l$ of S, $\beta \vdash S$, is a collection of subsets, or *blocks*, $B_1, \ldots, B_l$ whose disjoint union is S. The *type of* β is the integer partition

$$\lambda(\beta) = (|B_1|, |B_2|, \ldots, |B_l|),$$

where we assume that the B_i are listed in weakly decreasing order of size. ∎

For example, the partitions of $\{1, 2, 3, 4\}$ with two blocks are

$$1,2,3/4; \quad 1,2,4/3; \quad 1,3,4/2; \quad 2,3,4/1; \quad 1,2/3,4; \quad 1,3/2,4; \quad 1,4/2,3.$$

The first four are of type $(3, 1)$, while the last three are of type (2^2).

Let us first talk about the expansion of X_Γ in terms of monomial symmetric functions.

Definition 5.5.8 An *independent*, or *stable*, set of Γ is a subset $W \subseteq V$ such that there is no edge uv with both of its vertices in W. We call a partition $\beta \vdash V$ *independent* or *stable* if all of its blocks are. We let

$$i_\lambda = i_\lambda(V) = \text{the number of independent partitions of } V \text{ of type } \lambda. \quad ∎$$

Continuing with our example graph, the independent partitions of V are

$$v_1/v_2/v_3/v_4; \quad v_1, v_4/v_2/v_3; \quad v_2, v_4/v_1/v_3 \qquad (5.15)$$

having types (1^4) and $(2, 1^2)$. The fact that these are exactly the partitions occuring in the previous expansion of X_Γ into monomial symmetric functions is no accident. To state the actual result it will be convenient to associate with a partition $\lambda = (1^{m_1}, \ldots, d^{m_d})$ the integer

$$y_\lambda \overset{\text{def}}{=} m_1! m_2! \cdots m_d!$$

Proposition 5.5.9 ([Stn 95]) *The expansion of X_Γ in terms of monomial symmetric functions is*

$$X_\Gamma = \sum_{\lambda \vdash d} i_\lambda y_\lambda m_\lambda.$$

Proof. In any proper coloring, the set of all vertices of a given color form an independent set. So given $\kappa : V \to \mathbb{P}$ proper, the set of nonempty $\kappa^{-1}(i)$, $i \in \mathbb{P}$, form an independent partition $\beta \vdash V$. Thus the coefficient of $\mathbf{x}^\lambda$ in X_Γ is just the number of ways to choose β of type λ and then assign colors to the blocks so as to give this monomial. There are i_λ possiblities for the first step. Also, colors can be permuted among the blocks of a fixed size without changing $\mathbf{x}^\lambda$, which gives the factor of y_λ for the second. ∎

To illustrate this result in our example, the list (5.15) of independent partitions shows that

$$X_\Gamma = i_{(1^4)}y_{(1^4)}m_{(1^4)} + i_{(2,1^2)}y_{(2,1^2)}m_{(2,1^2)} = 1 \cdot 4!m_{(1^4)} + 2 \cdot 2!m_{(2,1^2)},$$

which agrees with our previous calculations.

We now turn to the expansion of X_Γ in terms of the power sum symmetric functions. If $F \subseteq E$ is a set of edges, it will be convenient to let F also stand for the subgraph of Γ with vertex set V and edge set F. Also, by the *components* of a graph Γ we will mean the topologically connected components. These components determine a partition $\beta(F)$ of the vertex set, whose type will be denoted by $\lambda(F)$. In our usual example, $F = \{e_1, e_2\}$ is a subgraph of Γ with two components. The corresponding partitions are $\beta(F) = v_1, v_2, v_3/v_4$ and $\lambda(F) = (3,1)$.

Theorem 5.5.10 ([Stn 95]) *We have*

$$X_\Gamma = \sum_{F \subseteq E} (-1)^{|F|} p_{\lambda(F)}.$$

Proof. Let $K(F)$ denote the set of all colorings of Γ that are monochromatic on the components of F (which will usually not be proper). If $\beta(F) = B_1/\ldots/B_l$, then we can compute the weight generating function for such colorings as

$$\sum_{\kappa \in K(F)} \mathbf{x}^\kappa = \prod_{i=1}^{l}(x_1^{|B_i|} + x_2^{|B_i|} + \cdots) = p_{\lambda(F)}(\mathbf{x}).$$

Now, for an arbitrary coloring κ let $E(\kappa)$ be the set of edges uv of Γ such that both $\kappa(u) = \kappa(v)$. Directly from the definitions, $\kappa \in K(F)$ if and only if $E(\kappa) \supseteq F$. So

$$\sum_{F \subseteq E}(-1)^{|F|}p_{\lambda(F)} = \sum_{F \subseteq E}(-1)^{|F|}\sum_{\kappa \in K(F)}\mathbf{x}^\kappa$$
$$= \sum_{\text{all }\kappa}\mathbf{x}^\kappa \sum_{F \subseteq E(\kappa)}(-1)^{|F|}.$$

But from Exercise 1e in Chapter 4, the inner sum is 0 unless $E(\kappa) = \emptyset$, in which case it is 1. Hence the only surviving terms are those corresponding to proper colorings. ∎

Returning to our running example, we can make the following chart, where the last line is the number of $F' \subseteq \Gamma$ with $|F'| = |F|$ and $\lambda(F') = \lambda(F)$ (including F itself):

F	$\emptyset$	e_1	e_1, e_2	e_1, e_4	e_1, e_2, e_3	e_1, e_2, e_4	E
$(-1)^{\|F\|}$	$+1$	-1	$+1$	$+1$	-1	-1	$+1$
$\lambda(F)$	(1^4)	$(2, 1^2)$	$(3, 1)$	(2^2)	$(3, 1)$	(4)	(4)
F'	1	4	5	1	1	3	1

Plugging this information into Theorem 5.5.10 we get

$$
\begin{aligned}
X_\Gamma &= p_{(1^4)} - 4p_{(2,1^2)} + 5p_{(3,1)} + p_{(2,2)} - p_{(3,1)} - 3p_{(4)} + p_{(4)} \\
&= p_{(1^4)} - 4p_{(2,1^2)} + 4p_{(3,1)} + p_{(2,2)} - 2p_{(4)}.
\end{aligned}
$$

The reader can check that this agrees with the earlier expansion in terms of monomial symmetric functions.

Many other beautiful results about X_Γ have been derived [Stn 95, Stn ta, S-S 93, Gas 96, G-S pr1, G-S pr2], but there are also a number of open problems, one of which we would like to state here. A *tree*, T, is a connected graph (only one component) that has no cycles, where a *cycle* is a sequence of distinct vertices $v_1, v_2, \ldots, v_n$ with $v_1 v_2, v_2 v_3, \ldots, v_n, v_1 \in E(T)$.

Proposition 5.5.11 *A tree T with $|V| = d$ has chromatic polynomial*

$$
P_T(n) = n(n-1)^{d-1}.
$$

Proof. Pick any $v \in V$ that can be colored in n ways. Since T has no cycles, each of the neighbors v can now be colored in $n - 1$ ways. The same can be said of the uncolored neighbors of the neighbors of v. Since T is connected, we will eventually color every vertex this way, yielding the formula for P_T. ∎

So all trees with d vertices have the same chromatic polynomial. But the situation appears to be the opposite for $X_T(\mathbf{x})$. To make this precise, we say that graphs Γ and Υ are *isomorphic* if there is a bijection $f : V(\Gamma) \to V(\Upsilon)$, called *an isomorphism*, such that $uv \in E(\Gamma)$ if and only if $f(u)f(v) \in E(\Upsilon)$.

Question 5.5.12 ([Stn 95]) *Let T and U be trees that are not isomorphic. Is it true that*

$$
X_T(\mathbf{x}) \neq X_U(\mathbf{x})? \; \blacksquare
$$

This question has been answered in the affirmative up through $|V| = 9$. However, we should note that it is possible to find a pair of nonisomorphic graphs that have the same chromatic symmetric function [Stn 95]. We hope that the reader will be enticed to work on this problem and the many others in the literature surrounding the symmetric group.

5.6   Exercises

1. Prove that the poset of partitions under dominance is always a lattice, but is not graded for $n \geq 7$.

2. Consider the *The Fibonacci poset*, Z, defined as follows. The elements of Z are all sequences w of 1's and 2's with covering relation $w \prec u$ if u is obtained from w by either

   Z1 deleting the first 1, or

   Z2 picking some 2 preceded only by 2's in w and changing it to a 1.

   (a) Prove that Z is differential.

   (b) Show that the number of elements of rank n is the *nth Fibonacci number*, defined by $F_0 = F_1 = 1$ and $F_n = F_{n-1} + F_{n-2}$ for $n \geq 2$.

3. Consider the poset X of all rooted trees τ (See Exercise 18 in Chapter 3) ordered by inclusion.

   (a) Show that X is a graded lattice .

   (b) Show that the nth rank has F_n elements (see the previous exercise for the definition of F_n).

   (c) Show that the saturated $\emptyset$–τ chains are in bijection with the natural labelings of τ.

   (d) Show that X is *not* differential, even though (by Exercise 18 in Chapter 3) the f^τ satisfy a sum-of-squares formula.

4. Prove Proposition 5.1.14.

5. Prove Theorem 5.1.15.

6. Let r be a positive integer. A poset is *r-differential* if it satisfies Definition 5.1.10 with DP2 replaced by the following:

   DP2r If $a \in A$ covers k elements for some k, then it is covered by $k + r$ elements.

   Prove the following statements about r-differential posets A.

   (a) The rank cardinalities $|A_n|$ are finite for all $n \geq 0$. (So the operators D and U are well-defined.)

   (b) Let A be a graded poset with A_n finite for all $n \geq 0$. Then A is r-differential if and only if $DU - UD = rI$.

   (c) In any r-differential poset

   $$\sum_{a \in A_n} (f^a)^2 = r^n n!,$$

   where f^a is the number of saturated $\emptyset$–a chains.

(d) If A is r-differential and B is s-differential, then the product $A \times B$ is $(r + s)$-differential. So if A is 1-differential, then the r-fold product A^r is r-differential.

7. Provide details for each of the three cases mentioned in Lemma 5.2.2.

8. In Theorem 5.2.4, fill in the proofs

    (a) of the case $k > i$,

    (b) that $Q_\pi = Q(\pi)$.

9. Come up with local rules for a growth $g'_\pi : C_n^2 \to Y$ such that the corresponding P'_π, Q'_π are the tableaux that would be those obtained by using column insertion in the Robinson-Schensted algorithm.

10. Use the results of this chapter to prove that $P(\pi^{-1}) = Q(\pi)$ and $Q(\pi^{-1}) = P(\pi)$.

11. Show that LR1–3 are a special case of DLR1–3. What is the corresponding function Ψ_μ?

12. (a) Prove Lemma 5.2.5.

    (b) Show that any growth $g : C_n^2 \to A$ with A differential must satisfy DLR1.

13. Prove Lemma 5.2.3 for any $g_\pi : C_n^2 \to A$ with A differential.

14. Fill in the details of Theorem 5.2.6.

15. Show that if g is an automorphism of a poset A, then g permutes the chains of any rank-selection A_S.

16. Prove (2) implies (1) in Theorem 5.3.3.

17. Let S be a set and let $S_1, S_2, \ldots, S_n$ be subsets of S. The set-theoretic Principle of Inclusion-Exclusion states that

$$|S \setminus \bigcup_{1 \le i \le n} S_i| = |S| - \sum_{1 \le i \le n} |S_i| + \sum_{1 \le i < j \le n} |S_i \cap S_j| - \cdots + (-1)^n |\bigcap_{1 \le i \le n} S_i|$$

Prove this in two ways: using a direct argument and by applying Theorem 5.3.3.

18. If A is a poset and $a \le b$ in A, then the corresponding *(closed) interval* is

$$[a, b] = \{c \in A : a \le c \le b\}.$$

Let Int A be the set of all closed intervals of A. Suppose that each interval of A is finite. In this case we say that A is *locally finite*. The

Möbius function of A is the function $\mu : \operatorname{Int} A \to \mathbb{Z}$ defined recursively by

$$\mu(a,b) = \begin{cases} 1 & \text{if } a = b, \\ -\sum_{a \le c < b} \mu(a,c) & \text{if } a < b, \end{cases} \tag{5.16}$$

where we write $\mu(a,b)$ for the more cumbersome $\mu([a,b])$. We also use $\mu_A(a,b)$ for $\mu(a,b)$ if it is necessary to specify the poset.

(a) Show that (5.16) uniquely defines μ.

(b) Show that (5.16) is equivalent to

$$\sum_{a \le c \le b} \mu(a,c) = \delta_{a,b}.$$

(c) Show that (5.16) is equivalent to

$$\sum_{a \le c \le b} \mu(c,b) = \delta_{a,b}.$$

(d) Posets A, A' are *isomorphic* if there is a bijection $f : A \to A'$ such that f and f^{-1} are both order preserving. Show that if A and A' are isomorphic and $f(a) = a'$, $f(b) = b'$, then $\mu_A(a,b) = \mu_{A'}(a',b')$.

(e) Prove that μ respects poset products in that

$$\mu_{A \times A'}((a,a'),(b,b')) = \mu_A(a,b)\mu_{A'}(a',b').$$

(f) Show that in the n-chain C_n we have

$$\mu(a,b) = \begin{cases} 1 & \text{if } a = b, \\ -1 & \text{if } a \prec b, \\ 0 & \text{otherwise.} \end{cases}$$

(g) Show that B_n is isomorphic to the n-fold product C_1^n, and so

$$\mu_{B_n}(S,T) = (-1)^{|T|-|S|}.$$

(h) Prove the Möbius Inversion Theorem: Given a vector space V, a locally finite poset A, and two functions $f, g : A \to V$ show that

$$f(a) = \sum_{b \le a} g(b) \;\; \forall a \in A \iff g(a) = \sum_{b \le a} \mu(b,a)f(b) \;\; \forall a \in A.$$

and

$$f(a) = \sum_{b \ge a} g(b) \;\; \forall a \in A \iff g(a) = \sum_{b \ge a} \mu(a,b)f(b) \;\; \forall a \in A$$

(i) Show that the Möbius Inversion Theorem implies Theorem 5.3.3.

19. Verify that equation (5.7) defines an action of $\mathcal{S}_n$ that is an automorphism of B_n.

20. Show that equation (5.8) defines an $\mathcal{S}_n$-isomorphism.

21. Construct an inverse for the map in Proposition 5.3.6.

22. Permutation $\pi = x_1 x_2 \ldots x_n$ has a *descent* at index i if $x_i > x_{i+1}$. The corresponding *descent set* is

$$\mathrm{Des}\,\pi = \{i \;:\; i \text{ is a descent of } \pi\}.$$

   (a) Show that if, by Robinson-Schensted, $Q(\pi) = Q$, then $\mathrm{Des}\,\pi = \mathrm{Des}\,Q$.

   (b) Let $\lambda \vdash n$, $S = \{n_1, n_2, \ldots, n_k\}_< \subseteq \{1, 2, \ldots, n-1\}$, and $\mu = (n_1, n_2 - n_1, \ldots, n - n_k)$. Then

   $$|\{\pi \in \mathcal{S}_n \;:\; Q(\pi) \text{ is a } \lambda\text{-tableau and } \mathrm{Des}\,\pi \subseteq S\}| = f^\lambda K_{\lambda\mu}.$$

   (c) For the action of $\mathcal{S}_n$ on B_n, decompose

   $$B^S \cong \sum_\lambda b^S(\lambda) S^\lambda.$$

   Then

   $$f^\lambda b^S(\lambda) = |\{\pi \in \mathcal{S}_n \;:\; Q(\pi) \text{ is a } \lambda\text{-tableau and } \mathrm{Des}\,\pi = S\}|.$$

23. Let $(a_k)_{k \geq 0} = a_0, a_1, a_2, \ldots$ be a sequence of positive real numbers.

   (a) Show that if the sequence of ratios $(a_k/a_{k+1})_{k \geq 0}$ is weakly increasing, then the original sequence is unimodal.

   (b) Show that if the original sequence is log concave (see Exercise 11 in Chapter 4), then it is unimodal.

24. Prove the second part of Lemma 5.4.4 in two ways: by mimicking the proof of the first part and as a corollary to the statement of the first part.

25. Give a different proof of Proposition 5.4.7 as follows. As before, it suffices to show that $\mathrm{rk}\,X_k = \binom{n}{k}$ for $k < n/2$. Let I be the $\binom{n}{k} \times \binom{n}{k}$ identity matrix.

   (a) Show that

   $$X_k X_k^t = X_{k-1}^t X_{k-1} + (n-k)I.$$

   Hint: Use the definition of X_k to calculate the (S, T) entry on both sides of the equation, where S, T are subsets of $\{1, 2, \ldots, n\}$ with k elements.

(b) Use the previous equation to show that $X_k X_k^t$ is positive definite, which implies that $\operatorname{rk} X_k = \binom{n}{k}$.

26. If $0 \le k \le l \le n$, let $X = X(k, l)$ be the matrix with rows (respectively, columns) indexed by the k-element (respectively, l-element) subsets of $\{1, 2, \ldots, n\}$ such that

$$X_{S,T} = \begin{cases} 1 & \text{if } S \subseteq T, \\ 0 & \text{otherwise.} \end{cases}$$

Note that $X_k = X(k, k+1)$. Show that if $k + l \le n$, then $X(k, l)$ has full rank in two ways: by mimicking the proof of Proposition 5.4.7 and by generalizing the previous exercise. For the latter you will need to prove the identity

$$X(k, l)X(k, l)^t = \sum_{i=0}^{k} \binom{n - 2k}{l - 2k + i} X(i, k)^t X(i, k),$$

which can be demonstrated using Vandermonde's convolution from Exercise 1f in Chapter 4.

27. Verify that $G \wr H$ is a group and that equation (5.14) does define an action for the wreath product.

28. A theorem of Whitney states that

$$P_\Gamma(n) = \sum_{F \subseteq E} (-1)^{|F|} n^{l(\lambda(F))},$$

where $l(\lambda(F))$ is the length of $\lambda(F)$. Prove this using results from the text. Note that as a corollary we have that $P_\Gamma(n)$ is a polynomial in n.

29. (a) Given an edge $e = uv \in E(\Gamma)$, let $\Gamma \setminus e$ be the graph obtained by deleting e from Γ. Also let Γ/e be the graph obtained by contracting e to a point. So in Γ/e, u and v merge into a new vertex x, and any edge wu or wv in Γ becomes an edge wx in Γ/e. (If both wu and wv exist, then they merge into a single edge wx.) Prove the Deletion-Contraction Law, which states that

$$P_\Gamma = P_{\Gamma \setminus e} - P_{\Gamma/e}.$$

Use this recursion to prove the following facts about P_Γ by induction on the number of edges:

(b) P_Γ is a polynomial in n.

(c) $\deg P_\Gamma = |V|$.

(d) The coefficients of P_Γ alternate in sign with leading coefficient 1.

(e) If Γ has k components, then the smallest power of n in P_Γ with nonzero coefficient is n^k.

Bibliography

[And 76] G. E. Andrews, "The Theory of Partitions," Encyclopedia of Mathematics and its Applications, Vol. 2, Addison-Wesley, Reading, MA, 1976.

[A-H 76] K. Appel and W. Haken, Every planar map is four colorable, *Bull. Amer. Math. Soc.* **82** (1976), 711–712.

[B-W 95] A. Björner and V. Welker, The homology of "k-equal" manifolds and related partition lattices, *Adv. in Math.* **110** (1995), 277–313.

[Boe 70] H. Boerner, "Representations of Groups with Special Consideration for the Needs of Modern Physics," North-Holland, New York, NY, 1970.

[C-H-R 86] A. R. Calderbank, P. Hanlon, and R. W. Robinson, Partitions into even and odd block size and some unusual characters of the symmetric groups, *Proc. London Math. Soc.* (1986), 288–320.

[C-R 66] C. W. Curtis and I. Reiner, "Representation Theory of Finite Groups and Associative Algebras," Pure and Applied Science, Vol. 11, Wiley-Interscience, New York, NY, 1966.

[DKR 78] J. Désarménien, J. P. S. Kung, and G.-C. Rota, Invariant theory, Young bitableaux and combinatorics, *Adv. in Math.* **27** (1978), 63–92.

[Dia 88] P. Diaconis, "Group Representations in Probability and Statistics," Institute of Mathematical Statistics, Lecture Notes-Monograph Series, Vol. 11, Hayward, CA, 1988.

[DFR 80] P. Doubilet, J. Fox, and G.-C. Rota, The elementary theory of the symmetric group, in "Combinatorics, Representation Theory and Statistical Methods in Groups," T. V. Narayama, R. M. Mathsen, and J. G. Williams, eds., Lecture Notes in Pure and Applied Mathematics, Vol. 57, Marcel Dekker, New York, NY, 1980, 31–65.

[Eul 48]   L. Euler, "Introductio in Analysis Infinitorum," Marcum-
            Michaelem Bousquet, Lausannae, 1748.

[Fom 86]   S. Fomin, Generalized Robinson-Schensted-Knuth correspon-
            dence, *Zapiski Nauchn. Sem. LOMI* **155** (1986), 156–175 [in Rus-
            sian].

[Fom 94]   S. Fomin, Duality of graded graphs, *J. Alg. Combin.* **3** (1994),
            357–404.

[Fom 95]   S. Fomin, Schensted algorithms for dual graded graphs, *J. Alg.
            Combin.* **4** (1995), 5–45.

[F-H 77]   W. Foody and A. Hedayat, On theory and applications of BIB
            designs with repeated blocks, *Ann. Stat.* **5** (1977), 932–945.

[FRT 54]   J. S. Frame, G. de B. Robinson, and R. M. Thrall, The hook
            graphs of the symmetric group, *Canad. J. Math.* **6** (1954), 316–
            325.

[F-Z 82]   D. S. Franzblau and D. Zeilberger, A bijective proof of the hook-
            length formula, *J. Algorithms* **3** (1982), 317–343.

[Fro 00]   G. Frobenius, Über die Charaktere der symmetrischen Gruppe,
            Preuss. Akad. Wiss. Sitz. (1900), 516–534.

[Fro 03]   G. Frobenius, Über die charakteristischen Einheiten der sym-
            metrischen Gruppe, *Preuss. Akad. Wiss. Sitz.* (1903), 328–358.

[Gan 78]   E. Gansner, "Matrix Correspondences and the Enumeration of
            Plane Partitions," Ph.D. thesis, M.I.T., 1978.

[G-M 81]   A. Garsia and S. Milne, A Rogers-Ramanujan bijection, *J. Com-
            bin. Theory Ser. A* **31** (1981), 289–339.

[Gas 96]   V. Gasharov, Incomparability graphs of $(3+1)$-free posets are s-
            positive, *Discrete Math.* **157** (1996), 193–197.

[G-S pr1]   D. Gebhard and B. Sagan, A noncommutative chromatic symmet-
            ric function, preprint.

[G-S pr2]   D. Gebhard and B. Sagan, Sinks in acyclic orientations of graphs,
            preprint.

[Ges um]   I. Gessel, Determinants and plane partitions, unpublished manu-
            script.

[G-V 85]   I. Gessel and G. Viennot, Binomial determinants, paths, and
            hooklength formulae, *Adv. in Math.* **58** (1985), 300–321.

[G-V ip]   I. Gessel and G. Viennot, Determinants, paths, and plane parti-
            tions, in preparation.

[G-J 83]   I. P. Goulden and D. M. Jackson, "Combinatorial Enumeration,"
           Wiley-Interscience, New York, NY, 1983.

[Gre 74]   C. Greene, An extension of Schensted's theorem, *Adv. in Math.*
           14 (1974), 254–265.

[GNW 79] C. Greene, A. Nijenhuis, and H. S. Wilf, A probabilistic proof of a
           formula for the number of Young tableaux of a given shape, *Adv.
           in Math.* **31** (1979), 104–109.

[Hai 92]   M. D. Haiman, Dual equivalence with applications, including a
           conjecture of Proctor, *Discrete Math.* **99** (1992), 79–113.

[Han 81]   P. Hanlon, The fixed-point partition lattices, *Pacific J. Math.* **96**
           (1981), 319–341.

[Her 64]   I. N. Herstein, "Topics in Algebra," Blaisdale, Waltham, MA,
           1964.

[H-G 76]   A. P. Hillman and R. M. Grassl, Reverse plane partitions and
           tableau hook numbers, *J. Combin. Theory Ser. A* **21** (1976), 216–
           221.

[Jac 41]   C. Jacobi, De functionibus alternantibus earumque divisione per
           productum e differentiis elementorum conflatum, *J. Reine Angew.
           Math. (Crelle)* **22** (1841), 360–371. Also in "Mathematische
           Werke," Vol. 3, Chelsea, New York, NY, 1969, 439–452.

[Jam 76]   G. D. James, The irreducible representations of the symmetric
           groups, *Bull. London Math. Soc.* **8** (1976), 229–232.

[Jam 78]   G. D. James, "The Representation Theory of Symmetric Groups,"
           Lecture Notes in Math., Vol. 682, Springer-Verlag, New York, NY,
           1978.

[J-K 81]   G. D. James and A. Kerber, "The Representation Theory of the
           Symmetric Group," Encyclopedia of Mathematics and its Appli-
           cations, Vol. 16, Addison-Wesley, Reading, MA, 1981.

[Kad pr]   K. Kadell, Schützenberger's "jeu de taquin" and plane partitions,
           preprint.

[Kar 88]   S. Karlin, Coincident probabilities and applications to combina-
           torics, *J. Appl. Probab.* **25A** (1988), 185–200.

[Kan 72]   W. M. Kantor, On Incidence matrices of finite projective and
           affine spaces, *Math. Z.* **124** (1972), 315–318.

[Knu 70]   D. E. Knuth, Permutations, matrices and generalized Young
           tableaux, *Pacific J. Math.* **34** (1970), 709–727.

[Kra pr]   C. Krattenthaler, Another involution principle-free bijective proof of Stanley's hook-content formula, preprint.

[Kre 65]   G. Kreweras, Sur une classe de problèmes de dénombrement liées au trellis des partitions des entiers, Cahiers du BURO, Vol. 6, 1965.

[Led 77]   W. Ledermann, "Introduction to Group Characters," Cambridge University Press, Cambridge, 1977.

[Lin 73]   B. Lindström, On the vector representation of induced matroids, Bull. London Math. Soc. 5 (1973), 85–90.

[Lit 50]   D. E. Littlewood, "The Theory of Group Characters," Oxford University Press, Oxford, 1950.

[L-R 34]   D. E. Littlewood and A. R. Richardson, Group characters and algebra, Philos. Trans. Roy. Soc. London, Ser. A 233 (1934), 99–142.

[Mac 79]   I. G. Macdonald, "Symmetric Functions and Hall polynomials," 2nd edition, Oxford University Press, Oxford, 1995.

[Mur 37]   F. D. Murnaghan, The characters of the symmetric group, Amer. J. Math. 59 (1937), 739–753.

[Nak 40]   T. Nakayama, On some modular properties of irreducible representations of a symmetric group I and II, Jap. J. Math. 17 (1940) 165–184, 411–423.

[NPS 97]   J. C. Novelli, I. M. Pak, and A. V. Stoyanovskii, A direct bijective proof of the hook-length formula, Discrete Math. Theoret. Computer Science 1 (1997) 53–67.

[PS 92]   I. M. Pak and A. V. Stoyanovskii, A bijective proof of the hook-length formula and its analogues, Funkt. Anal. Priloz. 26 (1992), 80–82. English translation in Funct. Anal. Appl. 26 (1992), 216–218.

[Pee 75]   M. H. Peel, Specht modules and the symmetric groups, J. Algebra 36 (1975), 88–97.

[Pra ta]   P. Pragacz, Algebro-geometric applications of Schur S- and Q-polynomials, Seminaire d'Algèbre Dubreil-Malliavin, to appear.

[PTW 83]   G. Pólya, R. E. Tarjan, and D. R. Woods, "Notes on Introductory Combinatorics," Birkhäuser, Boston, MA, 1983.

[Rem 82]   J. B. Remmel, Bijective proofs of formulae for the number of standard Young tableaux, Linear and Multilin. Alg. 11 (1982), 45–100.

[Rob 38]   G. de B. Robinson, On representations of the symmetric group, *Amer. J. Math.* **60** (1938), 745–760.

[Rob 91]   T. W. Roby, Applications and Extensions of Fomin's Generalization of the Robinson-Schensted Correspondence to Differential Posets, Ph.D. thesis, M.I.T., Cambridge, 1991.

[Rut 68]   D. E. Rutherford, "Substitutional Analysis," Hafner, New York, NY, 1968.

[Sag 80]   B. E. Sagan, On selecting a random shifted Young tableau, *J. Algorithms* **1** (1980), 213–234.

[Sag 82]   B. E. Sagan, Enumeration of partitions with hooklengths, *European J. Combin.* **3** (1982), 85–94.

[Sag 87]   B. E. Sagan, Shifted tableaux, Schur Q-functions, and a conjecture of R. Stanley, *J. Combin. Theory Ser. A* **45** (1987), 62–103.

[Sag 99]   B. E. Sagan, Why the characteristic polynomial factors, *Bull. Amer. Math. Soc.* **36** (1999), 113–134.

[S-S 90]   B. E. Sagan and R. P. Stanley, Robinson-Schensted algorithms for skew tableaux, *J. Combin. Theory Ser. A*, **55** (1990), 161–193.

[Sch 61]   C. Schensted, Longest increasing and decreasing subsequences, *Canad. J. Math.* **13** (1961), 179–191.

[Scu 01]   I. Schur, "Über eine Klasse von Matrizen die sich einer gegebenen Matrix zuordnen lassen," Inaugural-Dissertation, Berlin, 1901.

[Scü 63]   M. P. Schützenberger, Quelques remarques sur une construction de Schensted, *Math. Scand.* **12** (1963), 117–128.

[Scü 76]   M. P. Schützenberger, La correspondence de Robinson, in "Combinatoire et Représentation du Groupe Symétrique," D. Foata ed., Lecture Notes in Math., Vol. 579, Springer-Verlag, New York, NY, 1977, 59–135.

[Sta 50]   R. A. Staal, Star diagrams and the symmetric group, *Canad. J. Math.* **2** (1950), 79–92.

[Stn 71]   R. P. Stanley, "Ordered Structures and Partitions," Ph.D. thesis, Harvard University, 1971.

[Sta 75]   R. P. Stanley, The Fibonacci Lattice, *Fibonacci Quart.* **13** (1975), 215–232. MR **52** #7898

[Stn 82]   R. P. Stanley, Some aspects of groups acting on finite posets, *J. Combin. Theory Ser. A* **32** (1982), 132–161.

[Stn 88]   R. P. Stanley, Differential posets, *J. Amer. Math. Soc.* **1** (1988), 919–961.

[Stn 89]   R. P. Stanley, Log-concave and unimodal sequences in algebra, combinatorics, and geometry, in "Graph Theory and Its Applications: East and West," Ann. NY Acad. Sci. **576** (1989), 500–535.

[Stn 90]   R. P. Stanley, Variations on differential posets, in "Invariant Theory and Tableaux," D. Stanton ed., Springer-Verlag, New York, NY, 1990, 145–165.

[Stn 95]   R. P. Stanley, A symmetric function generalization of the chromatic polynomial of a graph, *Advances in Math.* **111** (1995), 166–194.

[Stn 97]   R. P. Stanley, "Enumerative Combinatorics, Vol. 1," Cambridge University Press, Cambridge, 1997.

[Stn 99]   R. P. Stanley, "Enumerative Combinatorics, Vol. 2," Cambridge University Press, Cambridge, 1999.

[Stn ta]   R. P. Stanley, Graph Colorings and related symmetric functions: ideas and applications, *Discrete Math.*, to appear.

[S-S 93]   R. P. Stanley and J. Stembridge, On immanants of Jacobi-Trudi matrices and permutations with restricted position, *J. Combin. Theory Ser. A* **62** (1993), 261–279.

[S-W 85]   D. W. Stanton and D. E. White, A Schensted algorithm for rim hook tableaux, *J. Combin. Theory Ser. A* **40** (1985), 211–247.

[Sun 94]   S. Sundaram, Applications of the Hopf trace formula to computing homology representations, *Contemp. Math.* **178** (1994), 277–309.

[S-Wa 97]   S. Sundaram and M. Wachs, The homology representations of the k-equal partition lattice, *Trans. Amer. Math. Soc.* **349** (1997) 935–954.

[S-We]   S. Sundaram and V. Welker, Group actions on linear subspace arrangements and applications to configuration spaces, *Trans. Amer. Math. Soc.* **349** (1997) 1389–1420.

[Tho 74]   G. P. Thomas, "Baxter Algebras and Schur Functions," Ph.D. thesis, University College of Wales, 1974.

[Tho 78]   G. P. Thomas, On Schensted's construction and the multiplication of Schur-functions, *Adv. in Math.* **30** (1978), 8–32.

[Tru 64]   N. Trudi, Intorno un determinante piu generale di quello che suol dirsi determinante delle radici di una equazione, ed alle funzioni simmetriche complete di queste radici, *Rend. Accad. Sci. Fis. Mat. Napoli* **3** (1864), 121-134. Also in *Giornale di Mat.* **2** (1864), 152–158 and 180–186.

[Vie 76]   G. Viennot, Une forme géométrique de la correspondance de Robinson-Schensted, in "Combinatoire et Représentation du Groupe Symétrique," D. Foata ed., Lecture Notes in Math., Vol. 579, Springer-Verlag, New York, NY, 1977, 29–58.

[Whi 84]   A. T. White, "Groups, Graphs, and Surfaces," North-Holland Mathematics Series, Vol. 8, New York, NY, 1988.

[Whi 83]   D. E. White, A bijection proving orthogonality of the characters of S_n, *Adv. in Math.* **50** (1983), 160–186.

[Why 32]   H. Whitney, A logical expansion in mathematics, *Bull. Amer. Math. Soc.* **38** (1932), 572–579.

[Wil 90]   H. S. Wilf, "Generatingfunctionology," Academic Press, Boston, MA, 1990.

[You 02]   A. Young, On quantitative substitutional analysis II, *Proc. London Math. Soc. (1)* **34** (1902), 361–397.

[You 27]   A. Young, On quantitative substitutional analysis III, *Proc. London Math. Soc. (2)* **28** (1927), 255–292.

[You 29]   A. Young, On quantitative substitutional analysis IV, *Proc. London Math. Soc. (2)* **31** (1929), 253–272.

[Zei 84]   D. Zeilberger, A short hook-lengths bijection inspired by the Greene-Nijenhuis-Wilf proof, *Discrete Math.* **51** (1984), 101–108.

Index

Action, 7
    on generalized tableaux, 79
    on standard tableau.x, 55
    on tabloids, 55
    orbit, 48, 210
    trivial, 11
Adjacent transposition, 4
Algebra, 4
    Boolean, 58, 192
    center of, 27
    commutant, 23
    endomorphism, 23
    full matrix, 4
    group, 8
    Weyl, 197
Algorithm
    backward slide, 113
    deletion, 94
    delta operator, 121
    dual Knuth, 172
    evacuation, 122
    forward slide, 113
    Greene-Nijenhuis-Wilf, 137
    Hillman-Grassl, 148–150
    insertion, 92
    Knuth, 169–171
    modified slide, 126–128
    Novelli-Pak-Stoyanovskii, 125–132
    Robinson-Schensted, 92–94
    straightening, 70
Alternant, 164
Ample, 209
Antidiagonal strip, 113
Appel-Haken, 213
Arm length of a hook, 125
Arm of a hook, 125

Automorphism, 204

Backward slide, 113
    modified, 128
Ballot sequence, 176
Block, 215
Boolean algebra, 58, 192
Boundary of a partition, 117
Branching rule, 77

Candidate cell, 128
Cauchy identity, 171
Cell, 54
    candidate, 128
Center, 27
    of commutant algebra, 27, 29
    of endomorphism algebra, 30
    of matrix algebra, 27
Centralizer, 3
Chain
    ascending, 194
    descending, 194
    saturated, 194
Character, 30
    defining, 31
    inner product, 34
    linear, 31
    orthogonality relations, 35
        first kind, 35
        second kind, 42
    product, 168
    regular, 31
    table, 32
Characteristic map, 167, 168
Chromatic
    polynomial, 214
    symmetric function, 214

Class function, 32
Code, 129
Coloring, 213
    proper, 213
Column insertion, 95
Column tabloid, 72
Column-stabilizer, 60
Commutant algebra, 23
Complement, 13
    orthogonal, 15
Complete homogeneous symmetric
       functions, 152, 154
Complete reducibility, 17
Components, 216
Composition, 67
Conjugacy
    class, 3
    of group elements, 3
Content, 78
Converge, 144
Corner, 76
Cover, 58
Cycle, 2
    in a graph, 217
    of length k, 2
    type, 2
Cyclic module, 57

Decreasing subsequence, 97
Degree
    of a monomial, 151
    of a power series, 144
    of a representation, 4
Deletion, 94
Delta operator, 121
Descent
    permutation, 221
    row, 70
    tableau, 206
Determinantal formula, 132
Differential poset, *see* Poset, dif-
       ferential
Dihedral group, 51
Direct sum, 13
Dominance lemma
    for partitions, 59

    for tableaux, 82
    for tabloids, 68
Dominance order
    for partitions, 58
    for tableaux, 82
    for tabloids, 68
Down operator
    in a differential poset, 196
    in Young's lattice, 193
Dual Cauchy identity, 172
Dual equivalence of tableau, 117
Dual Knuth equivalence, 111
    of tableaux, 113
Dual Knuth relations, 111

East, 129
Elementary symmetric functions, 152,
    154
Endomorphism algebra, 23
Equivalence, 99
    dual Knuth, 111
      of tableaux, 113
    dual tableau, 117
    Knuth, 100
      of tableaux, 113
    P, 99
    Q, 111
    tableaux, 114
Equivalent
    column tableaux, 72
    modules, 19
    row tableaux, 55
Euler, L., 144, 145
Evacuation, 122

Ferrers diagram, 54
Fibonacci
    number, 218
    poset, 218
Fixedpoint, 2
Fomin, S., 195, 197
Formal power series, 142
    convergence of, 144
    degree, 144
    homogeneous of degree n, 151
Forward slide, 113

modified, 126
Four Color Theorem, 214
Frame-Robinson-Thrall formula, 124
Frobenius reciprocity law, 48
Frobenius, G., 45, 166
Frobenius-Young formula, 132
Full matrix algebra, 4

Garnir element
    of a pair of sets, 70
    of a tableau, 71
General linear group
    of a vector space, 6
    of matrices, 4
Generalized permutation, 169
Generalized Young tableau, 78
Generating function, 142
    weight, 145
Gessel-Viennot, 158
Graded, 195
Grading, 152
Graph, 213
    chromatic polynomial, 214
    coloring, 213
    components, 216
    cycle, 217
    Four Color Theorem, 214
    independent partition, 215
    independent set, 215
    isomorphism, 217
    neighbors, 213
    proper coloring, 213
    stable partition, 215
    stable set, 215
    tree, 217
Greatest lower bound, 192
Greene, C., 102, 173
Greene-Nijenhuis-Wilf algorithm, 137
Group
    action, *see* Action
    algebra, 8
    character, 30
    cyclic, 5
    dihedral, 51
    general linear, 4, 6
    symmetric, 1

Growth, 197

Haiman, M. D., 117
Hasse diagram, 58
    normal, 112
    skew, 112
Hillman-Grassl algorithm, 148–150
Homogeneous of degree n, 151
Homomorphism, 18
    corresponding to a tableau, 80
Hook, 124
    arm, 125
    formula, 124
    leg, 125
    rim or skew, 180
    rooted tree, 138
    shifted, 139
    tableau, 126
Hooklength, 124
    rooted tree, 138
    shifted, 139

Inclusion-Exclusion, *see* Principle
    of Inclusion-Exclusion
Incomparable, 58
Increasing subsequence, 97
Independent
    partition, 215
    set, 215
Induced representation, 45
Inequivalent, 19
Inner corner, 76
Inner product
    invariant, 14
    of characters, 34
Insertion
    column, 95
    path, 92
    row, 92
Insertion tableau, 93
Interval of a poset, 219
Invariant
    inner product, 14
    subspace, 10
Involution, 2
Irreducibility, 12

Isomorphism
    graph, 217
    module, 19
    poset, 220

Jacobi-Trudi determinants, 158, 182
James, G. D., viii, 53, 63, 66, 182
Jeu de taquin, 116, 173
    backward slide, 113
    forward slide, 113
    modified backward slide, 128
    modified forward slide, 126
Join, 192

k-decreasing subsequence, 103
k-increasing subsequence, 103
Knuth, D. E.
    algorithm, 169–171
    dual algorithm, 172
    dual equivalence, 111
        of tableaux, 113
    dual relations, 111
        first kind, 111
        second kind, 111
    equivalence, 100
        of tableaux, 113
    relations, 99, 173
Kostka numbers, 85

Lattice
    path, 158
        involution, 161
        labelings, 158–159
        sign, 160
        weight, 159–160
    poset, 192
    Young, 192
        down operator, 193
        up operator, 193
Lattice permutation, 176
Least upper bound, 192
Ledermann, W., vii
Leg length
    of a hook, 125
    of a rim or skew hook, 180
Leg of a hook, 125

Lemma
    dominance, see Dominance lemma
    Schur's, 22
    sign, 64
Length, 152
Length of a subsequence, 98
Lexicographic order, 59
Littlewood-Richardson
    coefficient, 175
    rule, 177, 183
Local rules, 199
Locally finite, 219
Log concave, 189
Lower order ideal, 85

Möbius
    function, 219
    inversion, 206, 219
    inversion theorem, 220
Macdonald, I. G., viii, 164
Maschke's theorem
    for matrices, 17
    for modules, 16
Matrix, 4
    direct sum, 13
    tensor product, 25
Maximal element, 69
Maximum element, 69
Meet, 192
Miniature tableau, 118
Minimal element, 69
Minimum element, 69
Modified slide, 126–128
Module, 6
    complement, 13
    cyclic, 57
    direct sum, 13
    endomorphism algebra, 23
    equivalent, 19
    generated by a vector, 57
    homomorphism, 18
    inequivalent, 19
    irreducible, 12
    isomorphism, 19
    permutation module M^λ, 56
    reducible, 12

Specht, 62
tensor product, 44
unimodal sequence, 209
Monomial symmetric function, 151
Multiplicative, 154
Murnaghan-Nakayama rule, 180

n-chain, 58, 192
Natural labeling, 150
Neighbors, 213
Normal diagram, 112
North, 129
    step, 129, 158
Novelli-Pak-Stoyanovskii algorithm,
        125–132

Orbit, 48, 210
Order
    dominance, see Dominance or-
        der
    lexicographic, 59
    preserving, 197
Orthogonal complement, 15
Outer corner, 76

P-equivalence, 99
P-tableau, 93
Partial order, 58
Partial permutation, 101
Partial tableau, 92
Partially ordered set, see Poset
Partition, 2
    boundary, 117
    dominance
        lemma, 59
        order, 58
    hook, 124
    hooklength, 124
    length, 152
    normal, 112
    set, 215
    skew, 112
    strict, 139
Path
    Gessel-Viennot, 158
    Hillman-Grassl, 148

insertion, 92
    Novelli-Pak-Stoyanovskii, 127
    reverse Hillman-Grassl, 149
    reverse Novelli-Pak-Stoyanovskii,
        128
Permutation, 1
    adjacent transposition, 4
    cycle, see Cycle
    fixedpoint, 2
    generalized, 169
    involution, 2
    lattice, 176
    matrix, 6
    module M^λ, 56
    one-line notation, 2
    partial, 101
    representation, 8
    reversal, 97
    shadow diagram, 108
    shadow line, 107
    skeleton of, 110
    transposition, 4
        adjacent, 4
    two-line notation, 1
Permutation matrix, 6
Placement, 93
Polytabloid, 61
Poset, 58
    ample, 209
    ascending chain, 194
    automorphism, 204
    Boolean algebra, 58, 192
    chain, 194
    covering relation, 58
    descending chain, 194
    differential, 196
        down operator, 196
        Fibonacci, 218
        r-differential, 218
        up operator, 196
    graded, 195
    greatest lower bound, 192
    growth, 197
    Hasse diagram, 58
    incomparable elements, 58
    interval, 219

isomorphism, 220
join, 192
lattice, 192
least upper bound, 192
locally finite, 219
lower order ideal, 85
maximal element, 69
maximum element, 69
meet, 192
minimal element, 69
minimum element, 69
n-chain, 58, 192
natural labeling, 150
order preserving, 197
product, 197
r-differential, 218
rank, 195
rank-selected, 204
refinement of, 59
reverse partition, 150
saturated chain, 194
symmetric, 208
unimodal, 208
Power sum symmetric functions, 152, 154
Principle of Inclusion-Exclusion, 206, 219
Product of characters, 168
Product of posets, 197
Proper coloring, 213

Q-equivalence, 111
Q-tableau, 93

r-differential poset, 218
Rank, 195
Rank-selected poset, 204
Reciprocity law of Frobenius, 48
Recording tableau, 93
Reducibility, 12
    complete, 17
Refinement, 59
Representation, 4
    completely reducible, 17
    coset, 9, 20
    defining, 5, 8, 11, 14, 18, 20

degree, 4
direct sum, 13
induced, 45
irreducible, 12
matrix, 4
of cyclic groups, 5
permutation, 8
reducible, 12
regular, 8, 11
restricted, 45
sign, 5, 11, 19
tensor product, 43
trivial, 4, 11, 18
Young's natural, 74
Restricted representation, 45
Reversal, 97
Reverse ballot sequence or lattice permutation, 176
Reverse path
    code, 129
    Hillman-Grassl, 149
    Novelli-Pak-Stoyanovskii, 128
Reverse plane partition, 147
Rim hook, 180
    tableau, 184
Robinson-Schensted algorithm, 92–94
Roby, T. W., 197
Rooted tree, 138
    hook, 138
    hooklength, 138
Row word, 101
Row-stabilizer, 60
Rule
    branching, 77
    Littlewood-Richardson, 177, 183
    local, 199
    Murnaghan-Nakayama, 180
    Young, 85, 174

Saturated, 194
Schützenberger, M.-P., 106, 112, 121, 173, 177
Schur function, 155
    skew, 175
Schur's lemma

for matrices, 22
for modules, 22
over complex field, 23
Semistandard
basis, 82
Young tableau, 81
Sequence
symmetric, 208
unimodal, 208
Set partition, 215
block, 215
independent, 215
stable, 215
type, 215
Shadow, 107
diagram, 108
line, 107
x-coordinate, 108
y-coordinate, 108
Shape, 54
rooted tree, 138
shifted, 139
skew, 112
Shifted
hook, 139
hooklength, 139
shape, 139
Sign
of a permutation, 4
of a rim hook tableau, 185
of lattice paths, 160
representation, 5, 11, 19
Sign lemma, 64
Skeleton, 110
Skew
diagram, 112
hook, 180
Skew-symmetric functions, 163
Slide, 113
modified backward, 128
modified forward, 126
sequence, 114
South, 129
Specht module, 62
Stable
partition, 215

set, 215
Standard
basis, 67
Young tableau, 66
Stanley, R. P., 150, 195–197, 214
Step
eastward, 158
northward, 129, 158
westward, 129
Straightening algorithm, 70
Strict partition, 139
Submodule, 10
theorem, 65
Subsequence, 97
decreasing, 97
increasing, 97
k-decreasing, 103
k-increasing, 103
length, 98
Superstandard tableau, 157
Symmetric
group, 1
poset, 208
sequence, 208
Symmetric function
chromatic, 214
complete homogeneous, 152, 154
elementary, 152, 154
grading, 152
monomial, 151
multiplicative, 154
power sum, 152, 154
ring of, 151
Schur, 155
skew, 175

Tableau, 55
antidiagonal strip, 113
column-stabilizer, 60
content, 78
delta operator, 121
descent, 206
dominance
lemma, 82
order, 82
dual equivalence, 117

equivalent, 114
evacuation, 122
Garnir element, 71
generalized, 78
hook, 126
insertion tableau, 93
miniature, 118
of shape λ, 55
P-tableau, 93
partial, 92
Q-tableau, 93
recording tableau, 93
rim hook, 184
row descent, 70
row equivalent, 55
row word, 101
row-stabilizer, 60
semistandard, 81
standard, 66
superstandard, 157
type, 78
union, 120
Tabloid, 55
column, 72
dominance
lemma, 68
order, 68
of shape λ, 55
Tensor product
of matrices, 25
of representations, 43
of vector spaces, 26
Theorem
Möbius inversion, 220
Maschke
for matrices, 17
for modules, 16
submodule, 65
Thomas, G. P., 177
Total order, 58
Totally ordered set, 58
Transposition, 4
adjacent, 4
Tree, 217
Type
cycle, 2

set partition, 215
tableau, 78

Unimodal
module sequence, 209
poset, 208
sequence, 208
Union of tableaux, 120
Up operator
in a differential poset, 196
in Young's lattice, 193

Vandermonde
convolution, 186
determinant, 164
Vector space, 6
direct sum, 13
orthogonal complement, 15
tensor product, 26
Viennot, G., 106

Weakly
east, 129
north, 129
south, 129
west, 129
Weight, 145
generating function, 145
of lattice paths, 159–160
tableau, 155
West, 129
step, 129, 158
Weyl algebra, 197
Wreath product, 212

Young, 54
lattice, 192
down operator, 193
up operator, 193
natural representation, 74
polytabloid, 61
rule, 85, 174
subgroup, 54
tableau
generalized, 78
of shape λ, 55
semistandard, 81

standard, 66
tabloid, 20
   of shape λ, 55

Graduate Texts in Mathematics

(continued from page ii)

66 WATERHOUSE. Introduction to Affine Group Schemes.
67 SERRE. Local Fields.
68 WEIDMANN. Linear Operators in Hilbert Spaces.
69 LANG. Cyclotomic Fields II.
70 MASSEY. Singular Homology Theory.
71 FARKAS/KRA. Riemann Surfaces. 2nd ed.
72 STILLWELL. Classical Topology and Combinatorial Group Theory. 2nd ed.
73 HUNGERFORD. Algebra.
74 DAVENPORT. Multiplicative Number Theory. 3rd ed.
75 HOCHSCHILD. Basic Theory of Algebraic Groups and Lie Algebras.
76 IITAKA. Algebraic Geometry.
77 HECKE. Lectures on the Theory of Algebraic Numbers.
78 BURRIS/SANKAPPANAVAR. A Course in Universal Algebra.
79 WALTERS. An Introduction to Ergodic Theory.
80 ROBINSON. A Course in the Theory of Groups. 2nd ed.
81 FORSTER. Lectures on Riemann Surfaces.
82 BOTT/TU. Differential Forms in Algebraic Topology.
83 WASHINGTON. Introduction to Cyclotomic Fields. 2nd ed.
84 IRELAND/ROSEN. A Classical Introduction to Modern Number Theory. 2nd ed.
85 EDWARDS. Fourier Series. Vol. II. 2nd ed.
86 VAN LINT. Introduction to Coding Theory. 2nd ed.
87 BROWN. Cohomology of Groups.
88 PIERCE. Associative Algebras.
89 LANG. Introduction to Algebraic and Abelian Functions. 2nd ed.
90 BRØNDSTED. An Introduction to Convex Polytopes.
91 BEARDON. On the Geometry of Discrete Groups.
92 DIESTEL. Sequences and Series in Banach Spaces.
93 DUBROVIN/FOMENKO/NOVIKOV. Modern Geometry—Methods and Applications. Part I. 2nd ed.
94 WARNER. Foundations of Differentiable Manifolds and Lie Groups.
95 SHIRYAEV. Probability. 2nd ed.
96 CONWAY. A Course in Functional Analysis. 2nd ed.
97 KOBLITZ. Introduction to Elliptic Curves and Modular Forms. 2nd ed.
98 BRÖCKER/TOM DIECK. Representations of Compact Lie Groups.
99 GROVE/BENSON. Finite Reflection Groups. 2nd ed.

100 BERG/CHRISTENSEN/RESSEL. Harmonic Analysis on Semigroups: Theory of Positive Definite and Related Functions.
101 EDWARDS. Galois Theory.
102 VARADARAJAN. Lie Groups, Lie Algebras and Their Representations.
103 LANG. Complex Analysis. 3rd ed.
104 DUBROVIN/FOMENKO/NOVIKOV. Modern Geometry—Methods and Applications. Part II.
105 LANG. $SL_2(\mathbf{R})$.
106 SILVERMAN. The Arithmetic of Elliptic Curves.
107 OLVER. Applications of Lie Groups to Differential Equations. 2nd ed.
108 RANGE. Holomorphic Functions and Integral Representations in Several Complex Variables.
109 LEHTO. Univalent Functions and Teichmüller Spaces.
110 LANG. Algebraic Number Theory.
111 HUSEMÖLLER. Elliptic Curves.
112 LANG. Elliptic Functions.
113 KARATZAS/SHREVE. Brownian Motion and Stochastic Calculus. 2nd ed.
114 KOBLITZ. A Course in Number Theory and Cryptography. 2nd ed.
115 BERGER/GOSTIAUX. Differential Geometry: Manifolds, Curves, and Surfaces.
116 KELLEY/SRINIVASAN. Measure and Integral. Vol. I.
117 SERRE. Algebraic Groups and Class Fields.
118 PEDERSEN. Analysis Now.
119 ROTMAN. An Introduction to Algebraic Topology.
120 ZIEMER. Weakly Differentiable Functions: Sobolev Spaces and Functions of Bounded Variation.
121 LANG. Cyclotomic Fields I and II. Combined 2nd ed.
122 REMMERT. Theory of Complex Functions. *Readings in Mathematics*
123 EBBINGHAUS/HERMES et al. Numbers. *Readings in Mathematics*
124 DUBROVIN/FOMENKO/NOVIKOV. Modern Geometry—Methods and Applications. Part III.
125 BERENSTEIN/GAY. Complex Variables: An Introduction.
126 BOREL. Linear Algebraic Groups. 2nd ed.
127 MASSEY. A Basic Course in Algebraic Topology.
128 RAUCH. Partial Differential Equations.
129 FULTON/HARRIS. Representation Theory: A First Course. *Readings in Mathematics*
130 DODSON/POSTON. Tensor Geometry.

131 LAM. A First Course in Noncommutative Rings.
132 BEARDON. Iteration of Rational Functions.
133 HARRIS. Algebraic Geometry: A First Course.
134 ROMAN. Coding and Information Theory.
135 ROMAN. Advanced Linear Algebra.
136 ADKINS/WEINTRAUB. Algebra: An Approach via Module Theory.
137 AXLER/BOURDON/RAMEY. Harmonic Function Theory. 2nd ed.
138 COHEN. A Course in Computational Algebraic Number Theory.
139 BREDON. Topology and Geometry.
140 AUBIN. Optima and Equilibria. An Introduction to Nonlinear Analysis.
141 BECKER/WEISPFENNING/KREDEL. Gröbner Bases. A Computational Approach to Commutative Algebra.
142 LANG. Real and Functional Analysis. 3rd ed.
143 DOOB. Measure Theory.
144 DENNIS/FARB. Noncommutative Algebra.
145 VICK. Homology Theory. An Introduction to Algebraic Topology. 2nd ed.
146 BRIDGES. Computability: A Mathematical Sketchbook.
147 ROSENBERG. Algebraic K-Theory and Its Applications.
148 ROTMAN. An Introduction to the Theory of Groups. 4th ed.
149 RATCLIFFE. Foundations of Hyperbolic Manifolds.
150 EISENBUD. Commutative Algebra with a View Toward Algebraic Geometry.
151 SILVERMAN. Advanced Topics in the Arithmetic of Elliptic Curves.
152 ZIEGLER. Lectures on Polytopes.
153 FULTON. Algebraic Topology: A First Course.
154 BROWN/PEARCY. An Introduction to Analysis.
155 KASSEL. Quantum Groups.
156 KECHRIS. Classical Descriptive Set Theory.
157 MALLIAVIN. Integration and Probability.
158 ROMAN. Field Theory.
159 CONWAY. Functions of One Complex Variable II.
160 LANG. Differential and Riemannian Manifolds.
161 BORWEIN/ERDÉLYI. Polynomials and Polynomial Inequalities.
162 ALPERIN/BELL. Groups and Representations.
163 DIXON/MORTIMER. Permutation Groups.
164 NATHANSON. Additive Number Theory: The Classical Bases.
165 NATHANSON. Additive Number Theory: Inverse Problems and the Geometry of Sumsets.
166 SHARPE. Differential Geometry: Cartan's Generalization of Klein's Erlangen Program.
167 MORANDI. Field and Galois Theory.
168 EWALD. Combinatorial Convexity and Algebraic Geometry.
169 BHATIA. Matrix Analysis.
170 BREDON. Sheaf Theory. 2nd ed.
171 PETERSEN. Riemannian Geometry.
172 REMMERT. Classical Topics in Complex Function Theory.
173 DIESTEL. Graph Theory. 2nd ed.
174 BRIDGES. Foundations of Real and Abstract Analysis.
175 LICKORISH. An Introduction to Knot Theory.
176 LEE. Riemannian Manifolds.
177 NEWMAN. Analytic Number Theory.
178 CLARKE/LEDYAEV/STERN/WOLENSKI. Nonsmooth Analysis and Control Theory.
179 DOUGLAS. Banach Algebra Techniques in Operator Theory. 2nd ed.
180 SRIVASTAVA. A Course on Borel Sets.
181 KRESS. Numerical Analysis.
182 WALTER. Ordinary Differential Equations.
183 MEGGINSON. An Introduction to Banach Space Theory.
184 BOLLOBAS. Modern Graph Theory.
185 COX/LITTLE/O'SHEA. Using Algebraic Geometry.
186 RAMAKRISHNAN/VALENZA. Fourier Analysis on Number Fields.
187 HARRIS/MORRISON. Moduli of Curves.
188 GOLDBLATT. Lectures on the Hyperreals: An Introduction to Nonstandard Analysis.
189 LAM. Lectures on Modules and Rings.
190 ESMONDE/MURTY. Problems in Algebraic Number Theory.
191 LANG. Fundamentals of Differential Geometry.
192 HIRSCH/LACOMBE. Elements of Functional Analysis.
193 COHEN. Advanced Topics in Computational Number Theory.
194 ENGEL/NAGEL. One-Parameter Semigroups for Linear Evolution Equations.
195 NATHANSON. Elementary Methods in Number Theory.
196 OSBORNE. Basic Homological Algebra.
197 EISENBUD/HARRIS. The Geometry of Schemes.
198 ROBERT. A Course in p-adic Analysis.
199 HEDENMALM/KORENBLUM/ZHU. Theory of Bergman Spaces.
200 BAO/CHERN/SHEN. An Introduction to Riemann–Finsler Geometry.

201 HINDRY/SILVERMAN. Diophantine
    Geometry: An Introduction.
202 LEE. Introduction to Topological
    Manifolds.
203 SAGAN. The Symmetric Group:
    Representations, Combinatorial
    Algorithms, and Symmetric Functions.
    2nd ed.

204 ESCOFIER. Galois Theory.
205 FÉLIX/HALPERIN/THOMAS. Rational
    Homotopy Theory.
206 MURTY. Problems in Analytic Number
    Theory.
    Readings in Mathematics
207 GODSIL/ROYLE. Algebraic Graph Theory.